Best Practices for Transportation Agency Use of Social Media

Best Practices for Transportation Agency Use of Social Media

Edited by
Susan Bregman
Kari Edison Watkins, PhD

CRC Press
Taylor & Francis Group
Boca Raton   London   New York

CRC Press is an imprint of the
Taylor & Francis Group, an **informa** business

CRC Press
Taylor & Francis Group
6000 Broken Sound Parkway NW, Suite 300
Boca Raton, FL 33487-2742

Printed on acid-free paper
Version Date: 20130812

International Standard Book Number-13: 978-1-4665-6860-0 (Hardback)

Library of Congress Cataloging-in-Publication Data

Best practices for transportation agency use of social media / editors, Susan Bregman,
    Kari Edison Watkins.
        pages cm
    Includes bibliographical references and index.
    ISBN 978-1-4665-6860-0 (hardcover : alk. paper) 1. Transportation
    agencies--Customer services--United States. 2. Transportation agencies--Public
    relations--Public relations. 3. Social media--United States. 4. Internet in public
    administration-United States. I. Bregman, Susan. II. Watkins, Kari Edison.

HE203.B45 2014
354.76'23802854678--dc23   2013030832

Visit the Taylor & Francis Web site at
http://www.taylorandfrancis.com

and the CRC Press Web site at
http://www.crcpress.com

CONTENTS

EDITORS

Susan Bregman has more than 25 years of experience as a transportation researcher and policy analyst. She is the principal and founder of Oak Square Resources, LLC, a Boston-based consulting firm that provides research, policy, and communication services to the public transportation industry. She has particular expertise in the field of social media and was the principal investigator for TCRP Synthesis 99, *Uses of Social Media in Public Transportation*, published by the Transportation Research Board in 2012.

Since 2008, Susan has been editor of *The Transit Wire* (http://www.thetransitwire.com), a daily blog about transit technology that covers everything from mobile applications to social media to contactless fare collection. Before joining the consulting world, she worked for the City of Boston and the U.S. Department of Transportation. She started her professional life as a writer and editor for a trade association and a community newspaper.

Susan earned a master of city and regional planning degree from the Harvard Kennedy School and an undergraduate degree from Brown University with a specialization in linguistics. She is also an award-winning photographer and her work may be found at www.rednickel.com. Follow her on Twitter @OakSquareSusan.

Kari Edison Watkins, PE, PhD, is an assistant professor in civil and environmental engineering at Georgia Tech. After trying out the Northeast as a consultant for a decade and the Pacific Northwest to earn a PhD at the University of Washington (UW), she decided it was time to return to her undergraduate alma mater and the City of Atlanta to help it become a more transit-friendly, bikeable place.

In her research and teaching, Kari uses technology to improve, understand, and influence travel mode choice and multi-modal transportation planning. At UW, she co-created the OneBusAway program to provide real-time transit information tools and assess their impacts on riders in greater Seattle–Tacoma. She continues to work on ways to improve traveler information, but has also begun to examine opportunities to crowd-source cycling infrastructure and amenity data through the Cycle Atlanta program. Check out her research lab, UTIL, at http://watkins.ce.gatech.edu or follow her on Twitter @transitmom.

CONTRIBUTORS

Andrew Austin is the founding executive director of Americans for Transit. A political hack turned transit nerd, previously he was the field director at Transportation Choices, in Washington State, where he led numerous transit ballot measures and rider organizing campaigns. Prior to Transportation Choices, he managed political campaigns and was an aide in the Washington State legislature.

Jeremy Bertrand is the lead for digital outreach efforts at the Washington State Department of Transportation. He helped launch a social media program for WSDOT that has gained national acclaim for its efforts, receiving several national awards in the process.

Stacey Bricka, PhD, is a research scientist with the Texas A&M Transportation Institute (TTI). She has more than 20 years of experience in designing and conducting transportation-related surveys and analyzing travel behavior data. She views social media as an important tool for improving public participation in transportation surveys as well as in the transportation planning process itself. Stacey earned her doctorate in community and regional planning from the University of Texas at Austin.

Susit Dhakal is a master of public administration candidate at The University of Pennsylvania's Fels Institute of Government. Susit has worked in both for-profit and not-for-profit sectors in the United States and Nepal, his home country. As an associate consultant at Fels Research and Consulting, he is currently working with the Wyomissing Foundation to identify best practices across the United States in urban entrepreneurial development.

Jennifer Evans-Cowley, PhD, AICP, is the associate dean for academic affairs and administration for the College of Engineering and a professor of city and regional planning at The Ohio State University. In 2011, Planetizen.com, a public-interest information exchange for planners and designers, named Jennifer among the top 25 leading thinkers and innovators in the field of urban planning and technology, including her among

CEOs, developers, and academics around the world. She serves as a popular speaker on the importance of technology in cities. She earned a BS and master of urban planning from Texas A&M University and a master of public administration from the University of North Texas. Her PhD in urban and regional science is from Texas A&M University.

Tina Geiselbrecht is an associate research scientist for the Texas A&M Transportation Institute. Tina has extensive experience in public opinion research. She has facilitated over 100 focus groups ranging from strategic planning, to roadway safety, to innovative project delivery and finance. Her recent research focused on using social media to improve the transportation planning process. She earned a bachelor's degree from the University of Texas and a master's degree in geography and planning from Texas State University.

Samiul Hasan is a PhD candidate in transportation engineering and infrastructure systems at the School of Civil Engineering at Purdue University. He earned bachelor's and master's degrees from the Department of Civil Engineering of Bangladesh University of Engineering and Technology. His research interests include human mobility patterns and activity–travel analysis, urban analytics, machine learning, network modeling, hurricane evacuation, and agent-based simulation.

Ryan Hollingsworth is the communications and social media manager at Akron-Canton Airport (CAK) in Ohio. Ryan manages the airport's social media and communications including e-mail marketing, press materials, and website online materials. CAK was the first U.S. airport to join Facebook and is one of the top airport fan pages in the U.S., with more than 53,000 fans. CAK's social media has been recognized by MSNBC, *USA Today*, airport industry trade publications, and local media. In 2010, Likeable Media named Akron-Canton Airport one of the top 40 Facebook fan pages in the country. Ryan has spoken about social media at local and national conferences and frequently advises airports and businesses on best social media practices.

Katharine (Kate) Hunter-Zaworski, PhD, PE, is the director of the National Center for Accessible Transportation at Oregon State University. She is both a rehabilitation and transportation engineer. Her main research and development projects are related to improving access to intercity public transportation, including air and rail travel by people with mobility, sensory, and cognitive disabilities.

Steve Hymon is the editor of *The Source*, the official blog for the Los Angeles County Metropolitan Transportation Authority (Metro). He worked for 20 years in journalism, including stints at *Sports Illustrated* and the *Los Angeles Times*, where he was on a team of reporters who won the Pulitzer Prize for Public Service for a series of articles on a troubled county hospital.

Sarah M. Kaufman is a research associate at New York University's Rudin Center for Transportation, focusing on the use of cutting-edge information technologies in transportation communications, particularly the implementation of open data and social media programs. Sarah joined NYU after working for New York City Transit, where she served as a projects coordinator for emerging and intelligent transportation systems, developing customer communications technologies for mobile phones and in subways. Sarah earned a master's degree in urban planning from NYU Wagner and a master's in business administration-technology and innovation management from NYU's Polytechnic Institute. She earned a bachelor's degree in science writing with a focus on computer science from Washington University in St. Louis.

John Lisle is a former television journalist who transitioned to public affairs just when social media took off. He quickly realized its potential benefits for government communications and made the Arlington County (Virginia) Police Department a leader in the use of social media by launching a page on the most popular social networking site at the time, MySpace, and posting surveillance videos on YouTube to assist with criminal investigations. At the District Department of Transportation (DDOT) in Washington D.C., John created one of the most successful local government Twitter feeds @DDOTDC and managed the agency's portfolio of other media sites including Facebook, Google+, Flickr, YouTube, Scribd, and Pinterest. John is now the chief of external affairs at DC Water which also has a robust presence on the web including Twitter @DCWater.

Jody Feerst Litvak currently serves as community relations director for the Los Angeles County Metropolitan Transportation Authority (Metro). In her more than 20 years at the agency, she has also worked in planning, operations, and government relations. She especially thanks

xiii

her Metro colleagues Matthew Barrett and Lan-Chi Lam, along with outreach consultants at The Robert Group, for bringing her into the world of social media in 2007 for the Westside Subway Extension. A Los Angeles native, she earned degrees from the University of California Berkeley and Harvard University's Kennedy School of Government. She is looking forward to being able one day soon to ride the Westside Subway Extension and walk home.

Robin O'Hara is a director at the Los Angeles County Metropolitan Transportation Authority (Metro) and is responsible for front-end, customer-facing technologies in the digital fare card division. Robin has more than 20 years of marketing and communications experience and helped gain recognition for Metro with award-winning campaigns not only in transportation-based marketing contests, but also in general marketing competitions against such commercial giants as Mattel and Ford. She recently completed a master's degree in transportation management as valedictorian, with a capstone on social media.

Jamison Pack is the director of marketing with the Central Ohio Transit Authority (COTA) in Columbus. She is responsible for promoting COTA's service and identifies, develops, and manages business relationships. Jamison created and launched COTA's first social media plan, and created unique partnerships within the arts community, including Poetry in Motion, that can be enjoyed by bus riders today. Jamison came to COTA from WOSU Public Media in April 2011 where she worked as the assistant director of marketing and communications promoting NPR and PBS programming. She received an executive education certificate in marketing essentials from the Fisher College of Business and earned a bachelor's degree from The Ohio State University's College of Journalism and Communications.

Eric Rabe is an expert in communications and management strategies and is a senior advisor at the Fels Institute of Government, the public policy school at The University of Pennsylvania. Rabe writes and speaks on the intersection of management challenges and communications, and is an expert in new media strategies and tactics. His firm, Eric Rabe Communications Strategies, helps organizations analyze issues, plan long-term solutions, and execute effective in-house and external communications and issue management tactics. The firm's clients include senior executives of national and international organizations.

Ned Racine's 25 year's experience on large transportation projects covers Southern California and involves heavy rail, light rail, bus rapid transit, and highways. He currently serves as the Los Angeles County Metropolitan Transportation Authority's media officer for the I-405 Sepulveda Pass Improvements Project that will expand the nation's busiest stretch of highway. Unfortunately, he often drives the I-405 through the Sepulveda Pass.

Matt Raymond has marketed transportation in Denver, Dallas, and Los Angeles for the past 25 years and spent a decade as chief communications officer for Metro, the Los Angeles County Metropolitan Transportation Authority, where he pioneered social media in the public sector. Matt created Metro's blog, *The Source,* oversaw communications for "Carmageddon," and also initiated and led the effort to pass Measure R, a sales tax to support transportation in Los Angeles County. Matt introduced light rail in Denver and Dallas and launched more than two dozen major transportation projects across the country including the Eco Pass. Matt holds master's degrees in marketing, management, and public administration and a bachelor's in business from the University of Colorado and currently teaches transportation marketing and communications management at the Mineta Transportation Institute. Matt is the president and CEO of Celtis Ventures, LLC (CeltisVentures.com), a full-service venture marketing firm specializing in marketing, communications, and brand management located in Hermosa Beach, California.

Ashley Robbins is the campaign director for Americans for Transit, working in Georgia to organize riders across the state. Ashley's interest in transit led her to start her own transportation blog, finding her niche in the social media realm. An organizational psychologist, Ashley left consulting to work on transit issues, becoming president of Citizens for Progressive Transit and the social media coordinator and volunteer organizer for the Livable Communities Coalition's Transportation Investment Act campaign.

Nicole Sherbert is an assistant communication director for the Arizona Department of Transportation (ADOT) where she manages the agency's Office of Creative Services. Nicole's team is responsible for managing the ADOT brand identity, supporting community and media outreach efforts, and creating and delivering public safety and customer awareness campaigns through graphic design, video production, social media, web content, and campaign management. Prior to joining ADOT, Nicole worked in media relations for the City of Austin, Texas. She also managed strategic communications for various nonprofit organizations and political campaigns and spent several years as a reporter covering Texas politics. Nicole earned a master of public affairs degree from the LBJ School at the University of Texas and a bachelor's degree from the University of California at Davis.

Debbie Spillane is a researcher with Texas A&M Transportation Institute (TTI), focusing on analysis of survey data from the Texas statewide travel survey program and development of research proposals. Her research interests include social media and privacy issues related to survey technology. Spillane earned a BA from the University of Florida and a master's degree in sociology from Texas A&M University. Prior to joining TTI, she was a legal assistant for 10 years in law firms and private businesses.

Aaron Steinfeld, PhD, is an associate research professor at the Robotics Institute of Carnegie Mellon University. He specializes in operator assistance under constraints—that is, how to enable timely and appropriate interaction when technology use is restricted through design, tasks, environment, time pressures, and/or user abilities. His work focuses on intelligent transportation systems, human–robot interactions, rehabilitation, and universal design. He is the principal investigator and co-director of the Rehabilitation Engineering Research Center on Accessible Public Transportation (RERC-APT), the area lead for safe driving projects at the Quality of Life Technology Engineering Research Center (QoLT ERC), a project leader for technologies for the Safe and Efficient Transportation University Transportation Center (T-SET UTC), and a member of the NavLab research group. He is a co-founder of Tiramisu Transit, LLC, a spinout company of Carnegie Mellon University.

Timothy Tait, EdD, is an assistant communication director for the Arizona Department of Transportation (ADOT), leading the Office of Public Information. In this role, Tim oversees media relations, message development and positioning, constituent services, and the state's 511 Driver Information Center. Previously, Tim served as the agency's community relations director, managing public involvement and outreach efforts for the voter-approved, $9 billion Regional Transportation Plan freeway program for Metro Phoenix. Tim is a graduate of Arizona State University's Walter Cronkite School of Journalism. He earned a master's degree in crisis management and emergency response and a doctor of education degree in organizational management from Grand Canyon University. His dissertation examined the influence of specific leadership behaviors on the public involvement process. Before joining ADOT, Tim was a reporter for Phoenix area newspapers including *The Arizona Republic*, and served as the community affairs director for public school systems.

Anthony Tomasic, PhD, is a founder and the director of the Carnegie Mellon University's master of science in information technology (MSIT) program in very large information systems and a member of the Institute for Software Research (ISR) at the university. He has published in the areas of intelligent transportation systems, human–computer interactions, information retrieval, and databases. His research currently focuses on crowdsourcing, mixed-initiative systems, and applied machine learning. He is also a co-founder and the managing partner of Tiramisu Transit, LLC.

Satish V. Ukkusuri, PhD, is an associate professor in the School of Civil Engineering at Purdue University where he teaches courses in transportation systems and freight and logistics planning. He is also the director of the university's Interdisciplinary Transportation Modeling and Analytics Laboratory. His current areas of interest include dynamic network modeling, large scale data analytics, disaster management issues, and freight transportation and logistics. He has published extensively on these topics in peer-reviewed journals and presented papers at conferences. Satish's work is supported by various agencies including the National Science Foundation, the U.S. Department of Transportation, the New York Metropolitan Transportation Council (NYMTC), various state departments of transportation, the Global Policy Research Institute, and the Purdue Research Foundation. He is an area editor for *Networks and Spatial Economics* and an associate editor of *Transportmetrica Part B*. He is on the editorial advisory boards of *Transportation Research Part B* and *Transportation Research Part C*.

Kristie VanAuken is the senior vice president and chief marketing and communications officer at Akron-Canton Airport (CAK) in Ohio. She manages and directs all airport marketing and communication programs including air service development, branding, advertising, media relations, social media, and internal and external communications. Kristie is often invited to speak about airport marketing, branding, and social media at local, regional, and national conferences and forums. Kristie authors *It's a Trip*, a popular airport blog, and can be found on Twitter at @CAKristie. Prior to joining the airport in 1996, Kristie worked for the Greater Akron Chamber and the State of Michigan's International Affairs Division. In 2011, Kristie was recognized as a "Woman of Note" in *Crain's Cleveland Business Magazine*. She was also named one of the top 100 Women in Ohio Aviation by the International Women's Air and Space Museum in 2009.

Thomas Wall is a doctoral candidate in transportation systems engineering at the Georgia Institute of Technology, where he conducts research in the areas of infrastructure management and transportation planning. Specific research includes climate change adaptation planning and policy for transportation systems and the applications of social networking and media in transportation planning. He holds an Honors BS from Oregon State University and an MS from the Georgia Institute of Technology.

Sarah Windmiller is a graduate research assistant at the Georgia Institute of Technology. Her research concerns equity issues in mobile transit information and alternative solutions to improve the accessibility of this information. She is currently pursuing a master's degree in transportation systems engineering at the Georgia Institute of Technology, specializing in public transportation. Windmiller earned a bachelor's degree in urban planning and development from Ball State University in 2012.

Xianyuan Zhan is a PhD candidate in transportation engineering and infrastructure systems at Purdue University. Before entering graduate school, Xianyuan earned a bachelor's degree from the School of Civil Engineering at Tsinghua University in 2011. His research interests include large-scale data for traffic analytics, network modeling, and machine learning.

John Zimmerman is an associate professor at Carnegie Mellon University with joint appointments at the HCI Institute and at the School of Design. He conducts research in the areas of social, mobile, and ubiquitous computing, service design, user experience design, and interaction with intelligent systems. He teaches courses on interaction design and the design of mobile services.

1

Introduction

Social media constitutes a group of online applications that encourages users to create content and interact with one another. Examples include blogs, social and professional networking sites like Facebook and LinkedIn, micro-blogging sites such as Twitter, media-sharing sites like YouTube and Flickr, social aggregation applications like Pinterest, and location-based sites like Foursquare.

Sometimes called social networking or Web 2.0, social media provides government transportation agencies with unparalleled opportunities to connect with their customers. These connections may take many forms, but they all can help agencies engage with their stakeholders and personalize what can otherwise seem like a faceless bureaucracy. Transportation agencies have begun to adopt these networking tools, and their reasons for doing so typically fall into seven broad categories.

- Timely updates — Social media enables agencies to share real-time information and advisories including highway traffic conditions, construction-related detours, and transit service information. Some agencies also use social media to distribute emergency and/or weather-related communications covering parking restrictions, evacuation routes, and storm-related service delays.
- Public information — Many transportation organizations use social media to provide the public and the press with information about highway programs, transit routes and schedules, fares and tolls, and capital construction projects.
- Citizen engagement — The interactive nature of social media allows transportation organizations to build support for

programs and projects by connecting with citizens through informal online conversations.

- Marketing — Government agencies use social media to create brands and personalities for their organizations.
- Research — Transportation organizations use social media to learn about their customers by soliciting direct feedback or by analyzing the content of online posts.
- Customer service — Some transportation agencies encourage customers to use Twitter or other social media channels to ask questions or report problems such as broken air conditioning on train cars.
- Entertainment — Finally, social media can be fun. Agencies often use social media to display personal touches and entertain their constituents with photographs, videos, and contests.

Interactivity distinguishes social media from other static forms of communication, and these two-way communications provide a way for transportation organizations to listen and learn from their customers and stakeholders.

The purpose of this book is to provide guidance to transportation organizations interested in starting social media activities or modifying existing programs. Resources are provided throughout to help organizations understand aspects and uses of social media, recognize future trends and applications, and create policies and protocols related to social media activities. However, this is not a how-to guide. The authors assume that readers are already familiar with social media. The book does not include basic directions about setting up an account or similar guidance; this kind of information is readily available elsewhere.

Instead, the book incorporates contributions from government practitioners, academic researchers, and industry experts who talk about the best practices and lessons learned in social media usage among transportation agencies. Information about all modes of transportation including mass transit, highways, aviation, ferries, bicycling, and walking has been incorporated. Examples have been drawn from departments of transportation, public transportation operators, airports, nonprofit organizations, and advocacy groups.

The book begins in Chapter 2 where Susan Bregman of Oak Square Resources provides an overview of social media platforms and examples of usage of each platform among transportation organizations. For Chapter 3, Ms. Bregman conducted a series of interviews with some of the "best practice" agencies to better understand their use of social media and the lessons that can be learned. Interviewees include a regional airport (Akron-Canton), a city department of transportation (District of Columbia DOT), two state departments of transportation (Washington State and Arizona), and two transit agencies (Central Ohio Transit Authority and the Los Angeles County Metropolitan Transportation Authority).

Chapter 4, "Using Social Media to Connect with Customers and Community," contains four sections that explain more comprehensively how transportation organizations are using social media to inform and engage. Matt Raymond and Robin O'Hara, marketing experts with hands-on experience at transit agencies in Los Angeles, Dallas, and Denver, start with the use of social media in marketing to increase visibility and awareness. Ashley Robbins of the Livable Communities Coalition and Andrew Austin of Americans for Transit then discuss how social media can be used by advocacy organizations to build support. Next, Jody Litvak of the Los Angeles County Metropolitan Transportation Authority (Metro) and Jennifer Evans-Cowley of The Ohio State University discuss how social media can be used in public outreach and planning to obtain feedback from customers and commuters. Finally, Ned Racine, also of the Metro, joins Susan Bregman to describe how social media can be used as real-time communications tools.

Chapter 5, "Learning from Customers and Community," includes three sections that focus on more advanced uses of social media that enable transportation organizations to obtain information by involving customers and analyzing their use of social media in depth. First, Aaron Steinfeld, John Zimmerman, and Anthony Tomasic of Carnegie Mellon University discuss social computing, the combination of large groups of people and computing to accomplish projects that people and computers cannot complete alone. Stacey Bricka, Debbie Spillane, and Tina Geiselbrecht of the Texas Transportation Institute then join Tom Wall of Georgia Tech to discuss how transportation organizations use social media to recruit participants for market research surveys and supplement traditional survey efforts. Satish Ukkusuri, Samiul Hasan, and Xianyuan Zhan of Purdue University conclude with how researchers and transportation agencies can use social media analytical tools to identify service strengths and weaknesses and to assess customer sentiment.

Despite the stated advantages of using social networking, government organizations also face unique challenges including limited staff resources, regulatory requirements for record keeping, and the need to provide information in formats accessible to people with disabilities.

Chapter 6, "Agency Considerations and Policies," contains four sections to aid organizations in recognizing, understanding, and overcoming such barriers and concerns. In the first section, Sarah Kaufman of New York University's Rudin Center teams with Susan Bregman to discuss the resource requirements of maintaining a social media presence and policies to manage accounts and employee access; they also discuss archiving and records retention. Then Susan Bregman continues with a discussion about responding to online criticism and engaging customers and critics. Katharine Hunter-Zaworski of Oregon State University joins Kari Watkins and Sarah Windmiller of Georgia Tech to discuss how agencies using social media can equitably provide information to citizens despite the digital divide, language barriers, and accessibility challenges for patrons with disabilities. The chapter concludes with Eric Rabe and Susit Dhakal of the Fels Institute of the University of Pennsylvania, who discuss how agencies can measure the impacts of a social media presence.

Finally, Chapter 7 summarizes the lessons learned throughout the book and provides guidance to transportation organizations as they adapt to the changing social media landscape.

2

It's a Social World

Susan Bregman

Social media has revolutionized communications over the past decade. Only a few years ago, "friend" was not a verb and "tweeting" was for the birds. But now social networking has entered the mainstream and sometimes it seems like everyone—individuals, corporations, nonprofits, advocacy groups, and government agencies—has joined the conversation in the social space.

This chapter provides an overview of social media, describes some of the major social media platforms in use today, and explains how transportation agencies use these emerging communications tools.

WHAT IS SOCIAL MEDIA?

Social media, also called social networking, refers to a group of web-based applications that encourage users to interact with one another. Social media applications invite users to share opinions, information, experiences, photographs, and sometimes their locations.

Interactivity is one of the characteristics that distinguish social media from traditional media. Traditional media approaches such as press releases and websites tend to be centralized and focus on delivering a message to a specific audience. Social media methods are collaborative and rely on sharing information and soliciting feedback for their effectiveness. Traditional media techniques generally use one-way communication; social media tend to foster conversations. These connections can create a sense of engagement or loyalty among social media users.

Some of the best known social media platforms are Facebook, YouTube, Twitter, and LinkedIn, but hundreds, if not thousands of other social networks target specific interest groups.

WHO USES SOCIAL MEDIA?

Social media is truly an international phenomenon. In 2011, according to an analysis from comScore, social networking sites reached 82% of the world's online population or 1.2 billion users. That year, social networking accounted for nearly 19% of time spent online. Put another way, people around the world spent one of every five online minutes engaged in social networking (comScore 2011). Neilsen revealed similar findings for 2010, reporting that social networks and blogs accounted for 22% of time online, or one out of every four and a half minutes (NeilsenWire 2010).

In the U.S., the Pew Internet & American Life Project reported that two thirds of adult Internet users (67%) used a social networking site in 2012—more than double the 29% who reported using these platforms in 2008 (Duggan and Brenner 2013; Madden and Zickuhr 2011). ComScore (2011) reported an even higher figure, noting that 98% of online U.S. residents used social networking in 2011.

With the blurring line between personal and professional lives, many people access social media at the workplace for professional and personal reasons. Trend Micro (2010) found that 24% of U.S. corporate workers accessed social media sites while on the job in 2010—up from 20% in 2008. Another study of workers at U.S. corporations and not-for-profit organizations showed substantially higher use of social media in the workplace (SilkRoad 2012). Some 70% of surveyed workers used Twitter on the job for personal and professional purposes and 65% used Facebook. About 60% used their mobile devices to check social media activity more than once a day.

Seeing numbers like these, many transportation agencies have begun to incorporate social media into their approaches to marketing, communications, planning, and emergency management strategies. As this book demonstrates, reasons for doing so vary, but goals for using these communication channels may include disseminating information about services (including service disruptions and incidents), engaging with current riders, reaching out to potential riders, developing stronger community ties, enhancing the agency's branding and messaging, seeking feedback from stakeholders, and supporting customer service.

HOW DOES GOVERNMENT USE SOCIAL MEDIA?

Agencies and officials at all levels of government use social media, from city hall to the White House. For example, in a survey of more than 100 cities, Hansen-Flaschen and Parker (2012) found that the majority of surveyed municipalities used social media to connect with constituents. Most governments used Facebook (90%) or Twitter (94%) to maintain at least one department presence online, and nearly three of four cities had official Twitter accounts (74%) or Facebook pages (72%) set up by the mayor's office or communications department.

In addition to communications, these local governments use social media for activities like service improvement, public engagement, economic development, and emergency response. Governors have also adopted social media in a major way. Stateline reported that every U.S. governor had at least one social media account in 2011 (Mahling 2011). Some 47 governors (or their staffers) used Facebook and Twitter, 37 had YouTube accounts, and 27 used the photograph-sharing Flickr application.

At the federal level, the Government Accountability Office (GAO) reported that 23 of 24 major federal agencies used social media in 2011, specifically Facebook, Twitter, and YouTube (GAO 2011). According to the GAO, major reasons for using social media were to repost information from an agency's own website, to repost information from other sources, soliciting and responding to comments, and providing links to non-government websites.

HOW TRANSPORTATION ORGANIZATIONS USE SOCIAL MEDIA

Like other government agencies, transportation organizations have adopted social media over the past few years. In 2012, almost every state department of transportation had a social media presence, along with dozens of transit agencies, airports, turnpike authorities, and other transportation organizations at every level of government.

In its third annual survey of social media usage, the American Association of State Highway and Transportation Officials (AASHTO) showed that state departments of transportation were increasingly interested in improving the effectiveness of their social media programs (AASHTO 2012). Since 2010, when AASHTO began surveying state DOTs (and the District of Columbia) about their use of social media, overall adoption of these platforms has grown. Some other findings include the following:

- Twitter is the most popular social media platform among state DOTs; nearly three of four DOTs were using this application in 2012. Facebook ranked a close second.
- The use of the Flickr photo-sharing application has grown over the past few years, and 22 state-level DOTs were using the platform in 2012.
- While use of LinkedIn dropped between 2011 and 2012, DOTs began using two new social-curation platforms in 2012: Storify and Pinterest.

Figure 2.1 summarizes these trends.

The Transit Cooperative Research Program (TCRP) surveyed a small group of U.S. and Canadian transit agencies about their use of social media (Bregman 2012). In 2011, transit operators were most likely to use Twitter (91% of responding agencies), Facebook (89%), and YouTube (80%). Agencies

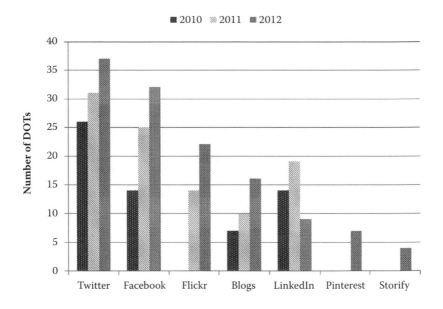

Figure 2.1 Use of social media by state departments of transportation. (Data adapted from: American Association of State Highway and Transportation Officials, 2012, 2011.)

were asked how they used social media; the most commonly stated uses were posting agency news, providing real-time alerts, and sponsoring contests and promotions. Transit providers were least likely to use social media for job posting and soliciting comments for public hearings.

Transit agencies did not adopt a one-size-fits-all approach to social media; instead, the survey responses suggested that organizations tried to take advantage of the unique characteristics of each social media platform. This was especially apparent with real-time service alerts. Three of four responding agencies (77%) used Twitter for these time-sensitive communications compared to only 49% that used Facebook for this purpose. Organizations preferred Twitter and Facebook for disseminating agency news, posting meeting and event notices, sponsoring contests and promotions, and distributing general service news. For feature stories, transit agencies opted for Facebook (57%) and blogs (40%). Only a few agencies reported using LinkedIn, mostly for job listings (14%) and service information (9%).

Airports have also turned to social media to connect with travelers; *SimpliFlying* reports that more than 200 airports around the world are active in the social space (Pal 2011). One particular challenge for airports is the "invisible wall" that is further described as the "seamless airport experience that leads most passengers to view the airport and airline as the same entity." Some airports use social media to help customers make the distinction between the airline providing a transportation service and the airport that provides a customer experience via retail shops, restaurants, ground transportation, and other features. Airports tend to prefer Twitter, especially for customer service, but the real-time application has also proven invaluable for crisis communications and weather emergencies.

Emerging travel services like bicycle- and car-sharing programs have been quick to adopt social media. The Hubway bicycle-sharing program in Boston, Massachusetts, for example, maintains an active Twitter account (@hubway) that engages program members in conversation like this recent exchange between Hubway and a member using the Twitter name @Lauren265:

> *@Lauren265:* Excellent first @Hubway night ride last evening. Boston drivers and fellow cyclists, beware!
> *@Hubway:* Glad to hear it! Hope you use the system again.

It is clear that governments at all levels, and transportation providers in particular, have adopted social media platforms for communicating with citizens, customers, and constituents. The next section describes the range of social media platforms in more detail and provides examples of how transportation agencies and organizations use these new communication tools.

DEFINING THE SOCIAL WEB

Social media experts have developed many organizational schemes to structure the social web.

The preferred taxonomy for this book includes the following categories:

- Social and professional networking sites that post information and encourage members to connect with one another
- Blogging sites where individuals or organizations post news, commentaries, photographs, and media clips
- Micro-blogging sites that share features with blogging and networking sites but limit posts to a very brief format
- Media- and document-sharing sites where members post and share video clips, photographs, documents, and presentations
- Social curation sites that encourage users to consolidate posts from multiple websites to illustrate or amplify a theme
- Geolocation applications that enable users to share their locations with other members of their social network and earn virtual "badges" for checking onto sites
- Crowdsourcing applications by which organizations tap into the collective intelligence of the public or a defined group to solve problems or prioritize issues, concerns, or recommendations

The list is not exhaustive but instead focuses on the types of social media platforms that transportation organizations—government and otherwise—are known to use. The following sections describe these social media platforms in greater detail and provide examples of how they are used by transportation agencies. Because of the constantly evolving nature of social media, this discussion should be considered a snapshot of the social media landscape in 2013.

Social Networking

Social networking sites are online platforms that support connections among members who may share personal or professional interests, activities, or backgrounds. Well known examples include Facebook, LinkedIn, and GooglePlus. For many people, social media start and end with sites like these.

Facebook

Facebook* is a social networking site that invites its users to create profiles, connect with other users, exchange messages, and share links, photographs, and video clips. Users can also set up groups around a particular topic or interest. Created in 2004, Facebook is one of the oldest social media platforms still in current use and arguably the best known around the world. As of December 2012, Facebook reported more than 1 billion monthly active users, of which about 82% were located outside the U.S. and Canada.

Facebook has two types of users: individuals and organizations. Individuals can create personal profiles, set up timelines to chronicle events in their lives, invite other users to be "friends," and share news and media clips. Personal posts and updates are typically visible only to members of an individual's network and are defined by a series of privacy settings. It should be noted that people have been known to set up personal Facebook pages for family pets or fake identities.

Organization users include businesses, government agencies, brands, public figures, and celebrities. They may set up profiles and provide updates for followers and constituents. Individuals may subscribe to an organization's updates and show their support by clicking a "Like" button. This is known as "liking" a page and replaced an earlier option to be a "fan" of a page. Unlike personal pages, anyone can see an institutional page—even people without Facebook accounts.

The following examples highlight how Amtrak, Zimride, and the Texas Department of Transportation—three very different transportation organizations—use Facebook.

With more than 285,000 likes (as of February 2013), Amtrak uses Facebook to provide information about its services and to engage its riders.† The railroad intersperses route descriptions with open-ended questions ("My travel plans during the month of September include ____.") and invitations to share photographs. Amtrak officials interact with customers, responding to comments and suggestions. For example, on September 30, 2012, Amtrak asked readers where they would be traveling in October. After nearly 200 riders responded with travel plans ranging

---

* https://www.facebook.com
† https://www.facebook.com/Amtrak

from Portland, Oregon, to Walnut Ridge, Arkansas, the railroad posted this reply: "We're so excited for all of these upcoming trips! Thank you all for traveling with us, and Happy October!"

Zimride is a social ride-sharing service that uses Facebook to connect drivers and passengers. The premise is simple. Drivers traveling within a specific corridor (such as Los Angeles to San Francisco) post information about their routes and schedules along with a suggested price per passenger. Potential passengers check out the listings and book a seat in a particular car. To help ensure security and encourage like-minded people to share rides, Zimride requires drivers and riders to access the system with a customized Facebook application that shares profile information between potential participants.[*]

The Texas Department of Transportation uses Facebook to inform citizens about department initiatives, circulate public safety guidelines, and provide updates on highway construction projects[†]. As of February 2013, TxDOT had more than 9,600 "likes" and posts ranged from the state's "Drive Clean across Texas" clean-air campaign and contest to the dangers of driving through flood waters. TxDOT's Facebook photography galleries highlighted workers as they conducted hurricane-readiness exercises, participated in community charitable events, and helped move a full-scale replica of the Shuttle Orbiter to a new home at Space Center Houston.

LinkedIn
LinkedIn[‡] is a professional networking site. The site launched in 2003 and by December 2012—less than 10 years later—the site had more than 200 million members. Like other social networking organizations, LinkedIn has a large international presence. According to the website, members are located in more than 200 countries and territories, and 64% of the membership is located outside the U.S.

LinkedIn members are encouraged to make connections with one another (the work-related equivalent of a Facebook "friend" relationship) and share professional highlights. A member can post an online profile including a detailed work history and resumé.

While LinkedIn is widely regarded as a networking site for individuals, companies and agencies also use the platform to share information, post links to relevant news stories and websites, and recruit employees.

---

[*] https://www.facebook.com/appcenter/carpool
[†] https://www.facebook.com/TxDOT
[‡] http://www.linkedin.com

When LinkedIn members follow organizations that have profiles on the site, they receive any updates posted by the organizations.

Organizational profiles are also intended to create professional networking opportunities. When an organization posts a profile on LinkedIn, users can easily identify individuals within the agency with whom they are connected, either directly or through one or more mutual contacts. For example, a LinkedIn member viewing the United States Department of Transportation's page would see a list of federal workers who are already in the member's network along with a list of department workers who attended his or her college. Another list would show all the DOT employees who are LinkedIn members regardless of their connections to the viewer. Organizations also have the option to post job listings on their pages.

LinkedIn also encourages users to participate in groups on a wide range of professional topics, where they can ask questions, share information, and forge new connections with colleagues in other organizations. LinkedIn has multiple groups of interest to transportation professionals, including groups geared toward traffic engineers, metropolitan planning officials, intelligent transportation systems, transportation consultants, young professionals, and university research centers.

GooglePlus

Launched in 2011, GooglePlus (also called Google+) is a social networking service owned and operated by Google.[*] The site reportedly had 250 million users in 2012, of which 150 million were active (O'Brien 2012). Like other social networks, GooglePlus allows members to create profiles, post updates and media, and follow other members. Because of its corporate parent, GooglePlus is integrated with other Google applications and services, including search results. The platform has not yet been widely adopted among transportation agencies. As of this writing only a few transportation organizations had accounts, including the Utah Department of Transportation, Caltrain, San Francisco Bay Area Rapid Transit (BART), and the District Department of Transportation in Washington, D.C.

Other Social Networking Sites

While Facebook and LinkedIn are arguably the best known social and professional networking sites, hundreds of other active and defunct social networks have been or are on the Internet. Some, like MySpace,[†] achieved

---

[*] https://plus.google.com
[†] https://www.myspace.com

13

great popularity and subsequently diminished in size and influence. MySpace began in 2003 and quickly became the go-to site for teenagers and young adults (Williams 2005). Until early 2008, the site was more popular than Facebook but its membership has waned (Arrington 2008). In 2012, MySpace reportedly had 25 million monthly unique visitors, far fewer than other networking sites (Ross 2012). Over the years the site shifted its focus from general social networking to music and is no longer relevant in the transportation world.

GovLoop* is a specialized networking site for government employees and those who work with government agencies. The site was founded in 2008 and has grown to more than 50,000 members. GovLoop encourages users to join discussions on topics ranging from government contracting to legal issues associated with social media to etiquette for job interviews via Skype. The site also offers resources for public employees at all levels, including training, job listings, and a federal per diem travel calculator.

GovLoop is built on architecture created by Ning† an online platform that enables users to create specialized social networks based around a particular interest, occupation, or connection. Creating a customized site for an online community can be a good solution when existing social networks do not fit the bill, and this approach is especially popular with advocacy groups and nonprofits.

Communities and citizen groups supporting goals of sustainable growth are well represented in the Ning universe. For example, Sustainable NE Seattle (sustainableneseattle.ning.com/) describes itself as a multi-neighborhood organization with action projects that include alternative transportation systems. Resources for members include an events calendar, a photo and video library, and interest groups like Local Motion which promotes bicycling, walking, and use of mass transit. Another group, called the New York State Paratransit Network (nysptn.ning.com/) established an online community of individuals and organizations that provide services to people with disabilities in the New York area.

Blogging

Derived from the term "web log," a blog is a regularly updated online journal. Blogs can be about any subject and usually contain comments from readers, photographs and media clips, and links to other websites.

---

* http:www.govloop.com
† http:www.ning.com

Blogs are often, but not exclusively, written by a single author and have no length restrictions. Posts are usually published in chronological order, but authors can tag posts with keywords and assign them to topics or categories so that users can easily locate information on a particular topic. Bloggers often use third-party tools like WordPress, Blogger, or Posterous to publish their sites.

Blogs can be a good choice for agencies that want to present detailed information about programs or policies, and several transportation organizations have used the platform with great success. The U.S. Transportation Security Administration (TSA), Los Angeles County Metropolitan Transportation Authority (Metro), and the U.S. Secretary of Transportation all use blogs to connect with constituents, as summarized below.

TSA uses a blog to showcase agency success stories and to explain policies. *The TSA Blog** runs a feature called "TSA Week in Review" in which the blogging team describes in great detail the range of illegal items confiscated at airport security checkpoints, from inert grenades to working firearms to an undeclared bottle of lotion. Figure 2.2 shows a sample entry. In a feature called "People Say the Darndest Things," the TSA bloggers recap comments from passengers that can get them in trouble, including "Would you help me get a bomb on a plane?" and the ever-popular "I have explosives in my bag." The blog uses the same informal tone throughout, even when admonishing travelers:

> Please, please, please, leave your grenades at home. Like milk and cola, grenades and airports do not mix, yet some still keep mixing them together. Please leave them at home or ship them via your preferred shipper (Burns 2012).

Los Angeles Metro introduced its blog known as *The Source†* in 2009. Instead of assigning editorial responsibilities to someone in the marketing department, the authority hired a Pulitzer Prize-winning journalist who previously worked at the *Los Angeles Times* as editor. Steve Hymon outlined his vision for the blog in its first issue:

> As a former newspaper reporter who covered this agency, my vision is to use *The Source* as a way to better explain and illuminate what's happening here and to do so in a way that isn't boring. I'll be editing *The Source* and mostly writing about policy and planning, i.e. projects that the agency is building, preparing to build and is dreaming of building (Hymon 2009).

---

* http://blog.tsa.gov
† http://thesource.metro.net

Figure 2.2 *The TSA Blog* presents a weekly summary of illegal or dangerous items discovered at security checkpoints. (Screen capture from *The TSA Blog* [blog.tsa.gov].)

Typical posts in *The Source* focus on Metro planning and construction projects, service updates, local events and promotions, board decisions, and feature stories about Los Angeles transportation history. The blog also features transit-related photographs submitted by readers, a summary of transit news from other properties, and a weekly round-up of tweets about Metro. Chapter 3 of this book contains an interview with *The Source's* editor Steve Hymon.

In 2011, Metro launched *El Pasajero*,* which was the first Spanish-language blog from a major U.S. transit agency. While the blog shares some content with *The Source*, the blog is not a direct translation and instead covers transportation stories affecting the region's Latino community.

---

* http://elpasajero.metro.net/

In addition to his YouTube videos, described elsewhere in this chapter, former Secretary of Transportation Ray LaHood was the voice of *Fast Lane* during his tenure. Described as the official blog of the U.S. Secretary of Transportation, *Fast Lane** offers updates about U.S. DOT programs and policies. Posts cover all modes of transportation and in any week the blog may talk about a visit to the St. Lawrence Seaway, present statistics on airline on-time performance, discuss federal transportation legislation, or offer a shout-out to DOT's summer interns.

Microblogging

Microblogging is considered a form of blogging where users post short updates and links to websites and online media. Twitter is the best known but not the only microblogging application.

Twitter

Twitter[†] is a real-time network that lets members share information in very short posts or "tweets." Each tweet is a maximum of 140 characters long (including spaces) and can include links to websites, photographs, and video clips. Unregistered users can read tweets, but only registered users can send them. Users can subscribe to or follow posts from other members, and members can share tweets with their followers by forwarding or "retweeting" updates of interest. Twitter user names are preceded by the @ symbol, such as "@Twitter." Twitter posts often include a "hashtag," which is a word or phrase preceded by a # symbol that functions as a searchable keyword.

Twitter claimed more than 200 million active users in February 2013. According to its website, it took Twitter 3 years, 2 months, and 1 day to reach its first 1 billion tweets. Now the platform sees about 340 million tweets daily or 1 billion every 3 days.

Government agencies and private businesses use Twitter to share information about their goods and services, collect feedback from customers and constituents, and engage with their followers. The real-time nature of Twitter makes it especially well suited for communicating time-sensitive information, from extreme-weather alerts to bus detours. Here is how some transportation organizations use Twitter to stay in touch.

---

* http://fastlane.dot.gov
† https://www.twitter.com

Figure 2.3 Twitter account for Huey P. Long Bridge shares update about its projected construction schedule. (From: Twitter account for Huey P. Long Bridge [@ hueypbridge].)

In Vancouver, BC, TransLink integrated Twitter into its customer service center.* This regional transit operator uses Twitter to monitor customer questions and complaints and respond in real time. Questions typically cover service frequency, fare zones, detours, and the like. On occasion, a customer has an unusual concern, like the time a rider wondered whether his monthly transit pass would still work after he accidentally ran his wallet through the wash. As of June 2013, TransLink's Twitter account had almost 33,000 followers.

Sometimes transportation agencies set up Twitter accounts for specific construction or planning projects; occasionally they give voice to a bridge or a train. The Louisiana Department of Transportation and Development set up social media accounts for a 7-year project to widen the Huey P. Long Bridge. Using the name @hueypbridge, the Twitter account primarily posts construction updates and lane closure information. See Figure 2.3.

In some cases, individuals set up unofficial (and occasionally outright fake) Twitter accounts for transportation services, including bridges, buses, and trains. The unofficial account for the 520 Bridge (@520_bridge) clearly states that its Twitter feed has no affiliation with the Washington State Department of Transportation; the tweets are a mix of retweeted

---

* https:www.twitter.com/translink

announcements from the department and bridge-related humor. An account purportedly representing the Hampton Roads Transit light-rail service known as *The Tide* (@NorfolkTide) appears to have no official status and offers a running commentary that can be witty, sarcastic, or both.

Not surprisingly, transportation agencies frequently take advantage of Twitter's real-time features to communicate time-sensitive information to their constituents. For example, the Indiana Department of Transportation (INDOT) has eight Twitter accounts. Six provide roadway information for the state's six geographic districts, one provides automated traffic updates from the state's traffic management center, and one provides general INDOT information. The New Jersey Transit Corporation (NJ TRANSIT) implemented a similar approach. In addition to @NJ_TRANSIT, which provides agency-wide service updates and customer service, the agency has separate Twitter accounts for each of its rail lines and two bus districts. This approach allows passengers on a specific commuter rail line to subscribe to route-specific updates without receiving irrelevant information about other transit services.

Twitter has also proven invaluable for emergency or crisis communications. For example, when a series of tornadoes struck Texas in 2012, Dallas/Fort Worth International Airport used Twitter to report cancelled flights and provide information for stranded travelers. "We had about 12 tornadoes touch down around #DFW yesterday. 100+ aircraft were damaged from hail," the airport tweeted from @dfwairport. "If you're in one of our terminals please don't hesitate to let us know if we can be of assistance" (Gaudin 2012).

Some municipalities and transportation departments established Twitter accounts specifically for posting emergency evacuation information in the case of a weather-related or other emergency. For example, the Texas Department of Transportation uses the Twitter handle @TxDOTalert for emergency-related information and the City of New Orleans uses @nolaready to tweet information about emergency preparedness.

Several transit agencies use Twitter for customer outreach, sponsoring online town halls at which the general manager and senior executives are available for a time to engage riders in two-way conversations. The Utah Transit Authority in Salt Lake City and TriMet in Portland, Oregon both sponsored such virtual meetings.

Tumblr
Tumblr* can be categorized as both a microblogging platform and a social curation site. The platform enables users to post text, photographs, music,

---

* https://www.tumblr.com

and videos to a personalized page called a Tumblog. As of August 2012, only a few government agencies were active on Tumblr and even fewer transportation organizations had accounts.

Among federal agencies, both the National Archives and U.S. Department of State use Tumblr. The National Archives shares one document from its collection each day via Tumblr.* Posts range from declassified memoranda on the Cuban missile crisis to nineteenth-century documents related to the purchase of Alaska. At a local level, the City of New York maintains an active Tumblr blog† to provide a user-friendly guide to city services along with a feature called "Ask Mike" that lets readers submit questions to Mayor Michael Bloomberg.

Two New York City transportation agencies also use Tumblr to stay in touch. The Department of Transportation manages an account called *The Daily Pothole*‡ that provides frequent updates about the number of potholes repaired and square miles of streets repaved (Figure 2.4). A link to an online form on its website allows citizens to report potholes to the department. The Metropolitan Transportation Authority's Arts in Transit and Urban Design program has a Tumblr blog called *Art along the Way* that features art installations throughout the New York City transit system.§

Media- and Document-Sharing Applications

Media-sharing sites allow users to upload and share different types of electronic media including video clips, photographs, reports, and presentations. YouTube and Flickr are among the best known applications for sharing videos and photographs, respectively, but social networks are available for sharing other types of media as well.

YouTube
Founded in 2005, YouTube is a video hosting site on which users can upload, watch, and share short videos.¶ Most of the content is generated by individuals, but companies, agencies, musicians, politicians, and organizations also share content on the site. Anyone can view a YouTube video, but only registered users may upload content. In 2013, YouTube boasted more than 800,000 unique users each month. Like Facebook, YouTube is an

---

* http://todaysdocument.tumblr.com/
† http://nycgov.tumblr.com/
‡ http://thedailypothole.tumblr.com/
§ http://artsfortransit.tumblr.com/
¶ http://www.youtube.com

Figure 2.4 New York City Department of Transportation maintains a Tumblog that posts a running count of the number of potholes repaired on city streets. (Screen capture from *The Daily Pothole*.)

international phenomenon; 70% of the site's traffic comes from outside the U.S. According to the site, about 72 hours of video are uploaded to YouTube every minute, and users watch over 4 billion hours of video each month.

Transportation agencies often use YouTube to educate and inform citizens about programs and policies. YouTube can also be an effective tool for building community support for a project or initiative. Here are some examples of how transportation agencies use YouTube.

The U.S. Department of Transportation uses YouTube to share public service advertisements for department initiatives like the *Faces of Distracted Driving* campaign which features interviews with people involved in accidents caused by distracted drivers or with their surviving family members.* In 2011, then Secretary of Transportation Ray LaHood started a monthly feature on YouTube called *On the Go* and he answered questions submitted by the public via Facebook, Twitter, and his blog. Typical topics included high-speed rail, livable communities, bicycle and

---

* http://www.youtube.com/user/usdotgov

Figure 2.5 U.S. Transportation Secretary uses YouTube to talk with citizens about transportation policy. (From: YouTube.)

pedestrian infrastructure, the benefits of carpools, and the impact of gas prices. Figure 2.5 shows the USDOT YouTube channel.

Dallas Area Rapid Transit (DART) used YouTube as part of an outreach program for its Green Line light rail project.* To build community support for the project during the 18-month construction period, media relations staff created a series of videos that showed how employees were connected to the Green Line project (Bregman 2012).

> For example, a DART police officer talked about her role in ensuring passenger safety, while a graphics designer highlighted his behind-the-scenes role in updating the system map. DART's goal was to post one update per week until the project was completed. By featuring its employees in the videos, especially those who normally did not work with the public, DART was able to "put a human face on a bunch of steel, concrete, and copper."

---

* http://www.youtube.com/dartdallas

22

After the project opened, DART continued to feature employees in YouTube videos. For example, it featured bus drivers in videos that showed riders how to make bus-to-rail connections at the new rail stations.

The Arizona Department of Transportation (ADOT) is one of many state transportation departments using YouTube to provide information about programs and projects. ADOT's videos cover construction projects, long-range planning, public service announcements, community outreach, and funding sources. The agency also has a Spanish-language channel.* Chapter 3 of this book includes an interview with ADOT's social media team.

Flickr

Flickr is a website that allows users to publish and share photographs (www.flickr.com). Members can upload photographs and videos to share with friends, family, and the world. Some members allow others to download their photographs under certain conditions and others retain all rights to their work. Flickr has a two-tier membership structure. General membership is free, but users also have the option to obtain professional accounts for monthly subscription fees.

Flickr was established in 2004 and acquired by Yahoo in 2005. Yahoo reports that Flickr has nearly 80 million unique global visitors, including 20 million in the U.S. About 4.5 million photographs are uploaded daily.

Many transportation agencies use Flickr to share photographs with customers, stakeholders, and the media. For example, the Metropolitan Transportation Authority in New York placed more than 2,000 photographs on Flickr that range from weekend track repairs to art installations at subway stations.† In the United Kingdom, Transport for London hosts its library of press images on Flickr.‡ The Transportation for America advocacy group uses Flickr to document good and bad examples of urban design.§

Other Media-Sharing Applications

Scribd¶ calls itself the world's largest online library and allows users to upload and share documents and reports. The District Department of Transportation in Washington, D.C., uses the platform to post planning studies and similar documents. Other media-sharing communities

---

* http://www.youtube.com/user/ArizonaDOT
† www.flickr.com/photos/mtaphotos/
‡ www.flickr.com/photos/tflpress/
§ www.flickr.com/photos/t4america/
¶ http//www.scribd.com

include SlideShare for presentations* and Vimeo for videos.† Streetfilms, an advocacy organization focusing on smart transportation design and policy, uses Vimeo to share short films that cover everything from transit applications for open data to the success of "The Porch," a pedestrian plaza installed next to Philadelphia's 30th Street Station.‡ Even the Google Drive online office tool (formerly Google Docs) incorporates a social component that allows users to edit and share documents in real time.

Social Curation Sites

While sites like YouTube and Flickr encourage people to share their own creations, social curation sites allow users to collect and display the work of others.

Pinterest

Pinterest is a fairly new curation application that lets users share (or "pin") photographs and images from online sources.§ The name combines "pin" and "interest" and is intended to suggest the bulletin boards or art boards commonly used in creative industries like fashion design or advertising. Pinterest started in 2010 as an invitation-only site and had 11.7 million unique users by January 2012 (Constine 2012).

While Pinterest began as a way for individuals to share images that amuse or inspire them, companies and organizations—including transportation agencies—have joined the site. For example, the Iowa Department of Transportation posted several boards or collections with images on various transportation-related themes, including "Bridges," "Safety," "Kids in Motion," and "Transportation Inspired Crafts."¶ The Rhode Island Department of Transportation posted an extensive collection of images on Pinterest, documenting everything from the Pawtucket River Bridge to Engineering Career Day.**

Storify

Launched in 2011, Storify allows users to create a story on a particular topic by assembling tweets, photos, videos, and other online content

---

* http://www.slideshare.net
† https://vimeo.com
‡ http://vimeo.com/streetfilms
§ https://pinterest.com
¶ https://pinterest.com/iowadot
** https://pinterest.com/ridotnews/

(storify.com). To provide context and background, users have the option to add their own comments and narratives. After creating their stories, users can share the final products with others on the Storify site or embed stories on their own web pages.

A few transportation organizations use Storify to connect with their communities. San Francisco Bay Area Rapid Transit (BART) pulled together tweets, photos, and videos to showcase new initiatives like a program that installed new vinyl seats on BART trains.[*] The Iowa Department of Transportation turned to Storify to create narratives about topics like motorcycle safety and teen drivers.[†]

LA Streetsblog, a transportation advocacy blog, uses Storify to document events and activities, especially when the group cannot provide firsthand accounts.[‡] Introducing a Storify stream about a new bicycle station in Long Beach, for example, *LA Streetsblog* (2011) writes, "When we can't make great events, we like to 'Storify' so that you can tell the story to us." The compilation from the Long Beach bike station used images from Flickr and posts from Facebook and Twitter to tell the story.

Geolocation Applications

Geolocation applications integrate location-specific information with data from smartphones and mobile devices enabled with geographic positioning systems (GPS) for two purposes: (1) to provide users with information about places and services in the vicinity, and (2) to allow users to share their locations with selected contacts. The latter feature is what takes an application into the social realm. Foursquare is the best known social application that relies primarily on geolocation, but Facebook, Twitter, and other social media platforms have also added geolocation capabilities.

Foursquare

Foursquare is a location-based web and mobile platform that enables users to "check in" to locations via smartphone application or text message (foursquare.com). Users also can share their location with friends by automatically posting updates on Facebook and Twitter. Foursquare was launched in 2009. As of January 2013, the platform claimed 30 million users worldwide and millions of check-ins daily.

---

[*] https://storify.com/sfbart/new-vinyl-seats-on-bart
[†] https://storify.com/Maverick
[‡] https://storify.com/lastreetsblog

Foursquare encourages participants to share their locations by awarding them points, virtual badges, and honorifics like "mayor" for checking into specific locations. Badges are linked to a specific number of check-ins at a location or type of venue. For example, users can access or "unlock" the virtual Trainspotter badge by logging a certain number of rail trips on any system. In addition, the individual with the most check-ins in the previous 60 days becomes "mayor" of that venue until ousted by another player.

To create a more rewarding experience for participants, Foursquare allows businesses and organizations to claim their locations such as retail stores and hotels (nicely blending the virtual and literal check-in experiences). Claiming a location enables an organization to gain access to analytical data on user check-ins, create special prizes and incentives for users, and post tips and updates. For example, a retail store might offer a $10 credit on a purchase or a restaurant may offer half-price appetizers to users who check into their locations.

Some transportation organizations—especially airports—have established their presence on Foursquare. Boston Logan International Airport manages a page and offers members "free BOS swag" when they achieve mayor status. The Norman A. Mineta San Jose International Airport offers passengers discounts from a changing array of airport shops when they check in. San Francisco's BART is one of the few public transportation agencies actively participating on Foursquare. BART set up Foursquare pages for its stations and encourages users to check in and share tips with other riders.* However, even if a transit agency has not claimed its stations or bus stops, individuals can still check into those locations informally.

Crowdsourcing

Crowdsourcing is a way for organizations to tap into the collective intelligence of the public or a defined group of customers or constituents to solve a problem or to prioritize issues, concerns, or recommendations. Like blogging, crowdsourcing is not tied into a specific vendor or application. Instead, multiple platforms are available for setting up an online conversation that allows users to submit suggestions, offer comments, and vote on their favorite ideas. Through this interactive process, the best ideas bubble up to the top.

The Utah Transit Authority devised the Next Stop Design project in cooperation with the University of Utah and the Federal Transit

---

* https://foursquare.com/sfbart

Figure 2.6 The Next Stop Design website showcases a winning crowdsourced bus stop design. (From: Next Stop Design. http://nextstopdesign.com/welcome. With permission.)

Administration (FTA) to design better bus stops via crowdsourcing.* Figure 2.6 shows the winning design.

In Spring 2012, Easter Seals Project ACTION teamed up with FTA and other federal agencies to sponsor an online dialogue about the transportation needs of veterans. The website used the IdeaScale feedback management crowdsourcing software to solicit comments from the public.[†] The New York City DOT used an online collaborative tool developed by OpenPlans to solicit suggestions for locating stations for the city's planned bicycle-sharing program.[‡]

---

* http://nextstopdesign.com
† http://veteransdialogue.ideascale.com/
‡ http://openplans.org/case-study/nyc-bike-share-map/

An increasingly common use of crowdsourcing is managing competitions for data-based products or mobile applications. For example, the U.S. Department of Transportation's Motorcoach Safety Data Student Challenge and New York City's BigApps competition were both run on the ChallengePost platform.*

Alternatively, organizations can use social media as crowdsourcing platforms. For example, an organization can structure the conversation on its Facebook page by asking open-ended questions or posting polls. Followers can vote, post comments, and support their favorite comments (using the Facebook "thumbs-up" feature). While this approach may not have all the features of a more formal crowdsourcing application, it allows organizations to quickly tap into the ideas of their followers.

OTHER SOCIAL MEDIA PLATFORMS

Social media are still evolving. This chapter presents a snapshot of the social space in 2013. While the social media platforms discussed in this chapter are some of the most commonly used, they are not by any means the only social services.

Already observers are seeing the impact of growth in tablet computers, smartphones and other mobile devices, third-party applications, location-based technologies, and social-buying services such as Groupon and LivingSocial. In fact, membership in Capital Bikeshare, the bicycle-sharing program in Washington, D.C., almost doubled overnight when the city partnered with LivingSocial to offer coupons for discounted membership (Schultz 2011).

SUMMARY

Transportation organizations are using multiple social media platforms to stay in touch with citizens and stakeholders. While experts have developed many organizational approaches to structuring the social web, this book uses the following categories: social and professional networking sites, blogging, microblogging, media- and document-sharing sites, social curation sites, geolocation applications, and crowdsourcing applications. Within these categories, the inventory of specific social media platforms keeps evolving and particular sites may gain or lose popularity over time.

---

* http://challengepost.com

Although the specifics may change, the interactive nature of social media sites has changed the way transportation organizations connect with their constituents.

REFERENCES

American Association of State Highway and Transportation Officials. 2011. Twitter, Facebook growing as effective media tools for state departments of transportation. Communications brief.

American Association of State Highway and Transportation Officials. 2012. *Third Annual State DOT Social Media Survey.*

Arrington, M. 2008. Facebook no longer the second largest social network. *TechCrunch.* http://techcrunch.com/2008/06/12/facebook-no-longer-the-second-largest-social-network/ (accessed October 1, 2012).

Bregman, S. 2012. *Uses of Social Media in Public Transportation.* TCRP Synthesis Report 99. Washington: Transportation Research Board.

Burns, B. 2012. TSA week in review: cornucopia of grenades. *TSA Blog.* http://blog.tsa.gov/2012/07/tsa-week-in-review-cornucopia-of.html (accessed August 14, 2012).

comScore. 2011. It's a social world: top 10 need-to-knows about social networking and where it's heading. *comScore*, December 21. http://www.comscore.com/Press_Events/Presentations_Whitepapers/2011/it_is_a_social_world_top_10_need-to-knows_about_social_networking.

Constine, J. 2012. Pinterest hits 10 million U.S. monthly uniques faster than any stand-alone site ever: comScore. *TechCrunch.* http://techcrunch.com/2012/02/07/pinterest-monthly-uniques/(accessed October 3, 2012).

Daily Pothole, The. (http://thedailypothole.tumblr.com/)

Duggan, M. and J. Brenner. 2013. The Demographics of Social Media Users, 2012. Pew Internet & American Life Project, February 13. http://pewinternet.org/~/media//Files/Reports/2013/PIP_SocialMediaUsers.pdf.

Gaudin, S. 2012. Facebook, Twitter are lifelines as tornadoes wallop Texas. *Computerworld*, April 4. http://www.computerworld.com/s/article/9225835/Facebook_Twitter_are_lifelines_as_tornadoes_wallop_Texas.

Government Accountability Office. Social Media: Federal Agencies Need Policies and Procedures for Managing and Protecting Information They Access and Disseminate. GAO 11-605. Report to Congressional Requestors. Washington, D.C.

Hansen-Flaschen, L. and K.P. Parker. 2012. The Rise of Social Government: An Advanced Guide and Review of Social Media's Role in Local Government Operations. Philadelphia: Fels Institute of Government of The University of Pennsylvania.

Hymon, S. 2009. Welcome to *The Source. The Source.* http://thesource.metro.net/2009/10/20/welcome-to-the-source/(accessed August 14, 2012).

LA Streetsblog. 2011. You tell the story: A new bike station opens in Long Beach. Storify. http://storify.com/lastreetsblog/you-tell-the-story-a-new-bikestation-opens-in-long (accessed August 23, 2012).

Madden, M. and K. Zickuhr. 2011. Sixty-Five Percent of Online Adults Use Social Networking Sites. Pew Internet & American Life Project, August 26. http://pewinternet.org/Reports/2011/Social-Networking-Sites.aspx

Mahling, M. 2011. How Many Governors Are Using Social Media? *Stateline* (blog), July 22. http://www.pewstates.org/projects/stateline/headlines/how-many-governors-are-using-social-media-85899376875.

NielsenWire. 2010. Social Networks/Blogs Now Account for One in Every Four and a Half Minutes Online. *NielsenWire* (blog), June 15. http://blog.nielsen.com/nielsenwire/global/social-media-accounts-for-22-percent-of-time-online/

O'Brien, T. 2012. Google+ has 250 million users, more mobile than desktop. *Engadget*. http://www.engadget.com/2012/06/27/google-has-250-million-users-more-mobile-than-desktop/ (accessed August 17, 2012).

Pal, S. 2011. Top 10 airports on social media: case studies of airports best at driving engagement. *SimpliFlying*, May 19. http://simpliflying.com/2011/top-10-airports-on-social-media-case-studies-of-the-airports-best-at-driving-engagement/

Ross, W. 2012. Justin Timberlake revives MySpace and the Internet reels. *The Daily Beast*. http://www.thedailybeast.com/articles/2012/09/28/justin-timberlake-revives-myspace-and-the-internet-reels.html (accessed October 1, 2012).

Schultz, D. 2011. Online coupon offer nearly doubles D.C. bike share membership overnight. *Transportation Nation*. http://transportationnation.org/2011/04/22/online-coupon-aides-in-d-cbikeshares-dramatic-ascent/ (accessed April 23, 2011).

SilkRoad. 2012. Social Media & Workplace Collaboration: 2012 Latest Practices, Key Findings, Hot Topics. http://pages.silkroad.com/rs/silkroad/images/Social-Media-Workplace-Collaboration-SilkRoad-TalentTalk-Report.pdf

The TSA Blog, http://blog.tsa.gov/2012/09/tsa-week-in-review-6-lbs-of-black.html)

Trend Micro. 2010. Corporate End User Study: Global Rise in Workplace Social Networking. July 12. http://trendmicro.mediaroom.com/index.php?s = 23 (accessed March 31, 2011).

Twitter, Huey P. Long Bridge, https://twitter.com/hueypbridge/status/298892033316245504)

Williams, A. 2005. Do You MySpace? *The New York Times*. http://www.nytimes.com/2005/08/28/fashion/sundaystyles/28MYSPACE.html?pagewanted = all&_r = 0 (accessed October 1, 2012).

Yahoo. 2011. *Flickr*. Yahoo Advertising Solutions. October 19. http://advertising.yahoo.com/article/flickr.html. (accessed August 17, 2012).

YouTube. USDOT. http://www.youtube.com/user/usdotgov?feature=CAQQ wRs%3D

3

Fish Heads and Haiku: Voices from the Field

Susan Bregman

To illustrate the opportunities and challenges of using social media on a day-to-day basis, co-editor Susan Bregman interviewed representatives from six transportation agencies across the U.S. The interviews highlighted common issues, successful practices, and lessons learned. The following agencies were included:

- Washington State Department of Transportation
- Los Angeles County Metropolitan Transportation Authority
- Akron-Canton Airport
- District Department of Transportation (Washington D.C.)
- Central Ohio Transit Authority
- Arizona Department of Transportation

These selected agencies include two transit operators, a regional airport, two state departments of transportation, and one city DOT. Each organization followed a different path to social media but they all share a common belief in the value of using these tools to provide information and connect with constituents.

ONE TWEET AWAY: ON SITE WITH THE WASHINGTON STATE DEPARTMENT OF TRANSPORTATION

Jeremy Bertrand is the web lead in the communications group at the Washington State Department of Transportation (WSDOT). The department has received accolades and awards for its use of social media including a blog, Twitter, Facebook, YouTube, and Flickr. WSDOT uses humor along with a few nods to popular culture to create memorable social media posts, some of which have received international attention.

How did WSDOT get started with social media?

We've been participating in some form of what is now called social media since 2006. We first came out with a blog to solicit feedback because of a winter storm. People were coming back from a Seahawks game and a sudden snowstorm hit harder than we had expected. People were very upset with us and thought that we should have done a better job. The Secretary of Transportation at the time was a very big fan of being transparent and accountable and 'fessing up and saying, "Hey, we made a mistake. How can we do better?" So the very first blog story we opened up was asking for feedback on how we did during that snowstorm.

Did you start a blog on the spot to do that?

On the spot. There were conversations: "Should we do it? Is it worth doing it? Who's going to staff it?" And then he said, "Hey, I want a blog set up first thing tomorrow morning." We contacted our IT department and they said, "Whoa, whoa, whoa. We've got to support this. What about our servers?" So, with their agreement, we took them out of the picture and used a third-party tool that they wouldn't have to support. We used Blogger at the time because it was one of the easiest ones to use.

Why did you start using Flickr?

Each phase was very purposeful. We opened a Flickr account because we were opening a second Tacoma Narrows Bridge in July 2007. We decided to let people walk on it before it was opened to traffic and take pictures and upload them to the web. The way we stumbled onto Flickr was we were thinking, "How in the world are we going to upload all these photos? It's going

32

to take someone a lot of time to resize them, optimize them for the web, and describe them." It was much easier to pay 25 bucks for a professional Flickr account and upload photos straight from the cameras from the bridge. It ended up being incredibly successful and it just went from there.

And Twitter?

Twitter came about because we saw such success with these other channels, and were starting to see a large audience in the Puget Sound area using the site. We started by putting our news releases there via RSS feeds. We didn't have a strategy yet; we just wanted to try it out and see what worked and what didn't. And then we started putting traffic alerts in there, which were just automated tweets.

When we started out we weren't personable, we weren't human. So it was really slow growth; there wasn't much of a following and we weren't creating any conversation. Then at a certain point, there was a snowstorm. We started providing real-time personal updates and people started responding back to us and our follower count jumped considerably. After that we completely reevaluated our use of it and instead of deciding the direction, we crowdsourced the audience and asked them what they thought we should do with the account.

So we asked users what we should do with these accounts and they said, "Stop doing the news releases and separate the traffic out to a different account because it's too much, and tell us what you're doing within the organization." Since then we've had a major storm event each year for the past 4–5 years, which has grown our Twitter account exponentially as people hunt and search for real-time information via Twitter.

It's become a matter of practice now within the agency to be the first source of transportation information. If we don't get there first we lose our opportunity to tell our story. We've even given cameras to all of our maintenance folks to take pictures of situations they encounter on the roadways. When they take a picture, they make sure that someone brings it back to the office for us to upload and share through social media channels. This has become a very successful practice because a photo of a road tells a better story than we could tell without it.

What sets WSDOT apart from other DOTs using social media?

There are several factors that set us apart. The first is our ability to relay information to the public quickly. We have a culture of being the first to provide information, good or bad, and that has really helped us to be the source of breaking transportation news in Washington State. Secondly, we were given the ability to post with a lot of autonomy and without an approval process from the start. This has allowed us to try a lot of different strategies, include spontaneous humor in the account while we continually reevaluate what is working and isn't working.

Some of our most famous content, which has gone international, is a combination of paying attention to the social zeitgeist and allowing employees to use cleverness to get messages out.

Your post about problems with Apple's new mobile mapping application got a lot of attention. Can you tell us what happened?

There was an iOS 6 update that forced iPhone users to use Apple's mapping software. [*Editor's note: With the release of the iOS 6 mobile operating system, Apple removed Google Maps from the iPhone 5 and introduced its own mapping application. Initial reviews found many flaws in the map information.*] So many people are on mobile devices now, and there are people who depend on mapping applications to make decisions about where to go. I wanted to tell people to be careful about using this new software in a tongue-in-cheek way. But rather than say, "Be careful about using iOS 6; it's not great," I took a picture, of the app, that showed the Tacoma Narrows bridges looking like one of the bridges had melted. And we said, "We can assure you that the Tacoma Narrows Bridge has not melted." (Figure 3.1) Then we used a hashtag on Twitter that everyone was using—I believe it was ios6—and that ended up getting national coverage with *Wired*, *The New York Times*, and CNET.

What have been your biggest challenges?

It's always easy to use social media during a storm. That's become standard now. The challenge is what to do between storms that will gain that audience and keep that content interesting enough that you're still relevant within the social media arena. There's a lot of competition within social media; you have to do something different to stand out to keep your audience interested.

Washington State DOT ✓
@wsdot

🐦 Follow

Although #ios6 may say differently, we can assure you that the Tacoma Narrows Bridges have not melted: bit.ly/NF6ccQ

← Reply   ⟲ Retweet   ★ Favorite

Oh no, Galloping Gertie is back!

t Tumblr @tumblr · Follow

714 RETWEETS   **119** FAVORITES

12:26 PM - 20 Sep 12 · Embed this Tweet

Flag media

Figure 3.1 Washington State DOT found a light-hearted way to draw attention to some of the issues associated with a new mobile mapping application. (From: Screen capture from WSDOT Twitter account.)

The other challenge was getting the rest of the agency to understand and see the importance of social media and incorporate it into their own communication plans. It's different from what we're used to saying in government communications. We're used to saying, "Be safe. Get ready for winter." What we're trying to do now is find that statement that will trigger an emotional response. What is that trigger that would cause a person to want to share a piece of content? What are those words/feelings that would make that content go further? So getting the internal audience to come around has taken a lot of work.

Who generates content now at WSDOT?

We have six or seven Twitter accounts right now. Two of them are automated. Then we have a central WSDOT account that one of the communications staff in the headquarters building manages. Often it's me, but there are several other people I pass it on to if I'm not here. Several communications people in the Seattle traffic management center monitor a Twitter account and a couple of people cover the account for the Vancouver office. So it's spread out within the agency.

Is there any centralized oversight of content or message?

The joke I say is that we're always one tweet away from being fired. So we have authority to post, but we do come together about once a week and are continually talking with each other to see if we could use different words that may have more success. We are in a constant state of reevaluation to learn from what we have done and take our content further. We have an 8 o'clock morning call for all the communications staff. We say, "What's happening today? What do we need to talk about? Who's going to be where? Who's going to be the lead on what?" And then we have a social media call about once a week. But even before we post something we often call one another and say, "What do you think? Is it going too far? Is it too inappropriate?" Sometimes it is, and we don't post it, sometimes we get a good clever laugh out of it, and sometimes we put it out anyway and get nervous that somebody will get in trouble for it.

So yes there is a lot of talking among ourselves because we've really pushed the limits of what a government agency can do in social media. And it's made it a fun and innovative place to work because of it. We've gotten great kudos because we pushed that limit a little bit. We're very fortunate that we're one of the few agencies with the authority to do that.

Do you use Flickr to disseminate photos to the media?

Yes, and the reason is we can get higher resolution on photos shared through Flickr than we can if we put it with a press release. I can put a 5 MB high-res photo that the *Seattle Times* can use for their print edition.

We will often put a whole set of photos with a story. Sometimes we won't even do a news release. We'll just put a set of photos up and just put stories beneath each one of the photos. And let the story tell itself.

What we've also done is take B-roll of major storm areas where the regular media can't get to and post it on YouTube. They'll actually play our YouTube footage in the background and use description of the clip in their story.

Who writes the tweets for the 520 Bridge?

The 520 Bridge has its own account (@520_bridge). One of the things we did recently was to take the time to go up and meet all the fake accounts and all the news accounts to say, "Hey we love what you're doing. How can we help?" And the 520 Bridge guy has been fantastic; we've had great conversations back and forth with him on our Twitter account. He's a big proponent of us because of that open conversation.

Do you have to archive posts and tweets?

We set an internal agency policy that we will archive 30 days' worth of posts. Honestly, I've yet to see a social media website remove any of the content that has been posted on it.

What sort of things do you measure?

A combination of things. For Twitter, retweets and mentions. For Facebook, comments and likes. For YouTube, it's views. Our in-house analytics tool is only focused on Twitter. Facebook and its API change so often, it's difficult to keep up with. But the measurement that has gotten us the furthest is the sharing of the emotional responses we receive from our followers of social tools, letting people within the agency see how useful our outreach is. One of the ways we do that is through the use

37

of Storify.* The use of this tool has been invaluable in letting a wider audience see the value.

Did you ever go back to your IT department and get their support?

No, because what we're doing has nothing to do with IT. The only challenge we had with IT was making sure we have the right browser so we can put this content out. Other than that, the tweets and Facebook posts and YouTube, it's all stuff that we've done within the communications office. Our IT office is just happy they didn't have to do anything with it. It didn't cost them anything extra, it didn't add any work for them, so they were all for it.

Do you have a social media policy?

Nothing official. We have a comment policy to say what we will and won't accept on social media channels.

We don't allow our employees access to social media channels due to state ethics rules and limited network bandwidth. Because they don't have access during the workday, we didn't feel the need to establish a full-on social media policy. We do have continuous conversations with those who manage our social media accounts, but keep access to social media locked down fairly tight.

What were some memorable social media stories?

We had a truck full of fish heads and guts and the truck tipped over and spilled all these fish guts over the roadway and closed the road. So we took a picture of that because we often try to say, "Here's why the road is closed." It's so much more interesting when there's a story behind it because it answers the "why." We were very fortunate that one of the maintenance guys had a camera, and we posted a photo of these fish guts all over the road (Figure 3.2). We cracked a couple of jokes about it, "It started off as a slow Friday and then this happened." And then we were having a conversation about it in our office, and somebody said, "Can you just believe that somebody microwaved fish in the break room?" So I threw that out as a comment, "Yeah, it gets worse. Someone just microwaved fish in the

---

* http://storify.com/wsdot

Figure 3.2 Fish heads and guts spill on U.S. 101. (From: Washington State Department of Transportation via Flickr, licensed under Creative Commons Attribution-NonCommercial-NoDerivs 2.0 Generic [CC BY-NC-ND 2.0]. With permission.)

break room so now we know what it smells like." That comment just made the post go further.

But you got lucky because of what was in the truck, didn't you? You can't plan those things.

It depends. A woman posted a note on our Facebook wall and said that her granddaughter lost her teddy bear when the car pulled over on I-90:

Lost Daddy Bear. Our 6-year-old granddaughter got car sick west of Ellensburg Tuesday, October 11. When we pulled over to the side of the road to help her, Daddy Bear, who is about 12 inches long and white fell out of the car …. Daddy is in the Army stationed in South Korea and Daddy Bear means the world to her.[*]

---

[*] http://www.facebook.com/WSDOT/posts/10150318383611975

We did some strategic thinking on how we could do that. A couple of our guys went out and found the bear and then brought it back to her. And it ended up being covered on *Good Morning America,* and Diane Sawyer covered it as the "Hero of the Day." You wouldn't believe how many news trucks were parked in front of that little girl's house when we brought her bear back to her. So there's a few where you know it's a good enough story that it's going to go somewhere. But yeah, sometimes you get lucky, but you have to make the most of it.

How do you handle online criticism?

We get several different types of criticism. There's the one guy who's not happy no matter what we do. He bases all of his attitudes and behaviors on what media articles come out about us, however true or not, and sometimes you just can't dig yourself out of that. However we do get a lot of the "F- You, WSDOT" comments and we'll actually reply to those and say, "Why are you so upset, man?" One guy actually deleted his post after we responded and said, "Sorry, it was one in the morning and you closed down the roadway to do work, and I just wanted to get home."

A lot of those, I would say 75% of them, are just emotional outbursts. We respond to most of them and say, "Hey, what happened? Sorry, we'll do our best, or we'll try next time, or we'll do this differently," or "Hey, here's how you can avoid this type of situation in the future." Most of them have come back and said, "Wow, thanks for the response." There are still some people who are absolutely blown away because a government agency is being responsive on these accounts.

One of our biggest critics commented on every single blog story that we put out for the first two years. And at one point we declined his comment because he used profanity. But instead of just declining it, we said, "Here's our number. Why don't you give us a call; we'd love to have a conversation with you about this." He now calls me several times a week to ask questions about things or to talk to me about state government in general. And it turns out he's just kind of a policy wonk who is disappointed in a lot of state government decisions and thinks they can do it better. He doesn't know how to communicate very well; he's just really gruff and sharp and pokes at you. But now that we've established a rapport it's almost a friendship now. We've developed a relationship with him and can have a real conversation about situations surrounding decisions that we made. That's a much easier conversation to have than having that guy post negative comments all the time. So you can reach some people. A few just don't like government and never will.

How do you measure success?

Our goal two years ago was to get a tweet retweeted more than 50 times. We were trying to tell people to get ready for a winter storm and I put a tweet out saying, "We're preparing for the worst storm in 30 years and we're hoping you are too." And I believe that got 60 retweets.

And then we had the geese story. There were some Canada geese on the road and a patrol had two cars trying to guide the geese off the roadway because they could cause an accident. We took video footage of that and put it on YouTube. And that went international. That's the first time I had to Google Translate Croatian and Brazilian comments on the YouTube site.

We had the teddy bear rescue which was on *Good Morning America* and Diane Sawyer. But that iOS6 story recently went at least national. It was covered by the tech audience and retweeted over 600 times. It was just amazing.

So we have many different types of success stories and not one that's more successful than another. And most of them are not necessarily the standard agency messaging, but they're clever stories that we encounter.

Our biggest success was taking the time to network with all the local news stations and meet personally their social media people who upload content. They're looking for a story just as much as anybody else is. And if we can share a story that helps them put something out, they're constantly calling us now or sending us texts saying, "Hey what about this?" or "We're doing a story about that." I think that's helped share our story more than anything else. That networking has paid off for me quite a bit. Now they're writing stories without even contacting us; they're just quoting our social media feeds.

41

Any mistakes?

I wouldn't have started out with newsfeeds on our Twitter account. I would have started out being a person from the beginning. I would have paid more attention to video earlier on. It's something we are still struggling with; people just aren't interested in government videos. Our video strategy is still not a direct strategy it's more reactionary and "hey, if somebody has time can we get this?" I would have made video a more focused conscious approach.

What we don't have is a social media person. My role now is both managing the website, internal and external, and monitoring the social media. I wouldn't wish that on anybody; it's too much for anyone to do. I wish we would have had a direct person responsible for social media who could write blog stories regularly and could take videos regularly and have a more regular calendar. I would have gotten more people in the agency involved earlier and gotten them up to speed sooner.

Any advice for other DOTs?

I would highly recommend they pay attention to what's happening out there—that pop culture zeitgeist. I think there is so much noise out there I feel bad for those agencies that are just now getting onto social media. On the other hand, there is so much knowledge out there that they really need to pay attention to that knowledge. There is no reason they shouldn't be able to succeed based on what's out there by watching how these companies succeed or fail based on what they're putting out.

I wish more of them really took the time to evaluate how successful they are with their social strategies. Just linking your incident feed up to a Twitter account can only take you so far. There is such an amazing return to be realized by humanizing the social account and it has great returns both in public interaction but also in education.

So jump out there and do it, but don't be afraid to change. Just like the Internet, the atmosphere and use of social media is constantly changing too as it matures.

KEEPING IT CONVERSATIONAL:
ON SITE WITH L.A. METRO

Steve Hymon edits *The Source* which is a blog for the Los Angeles County Metropolitan Transportation Authority (Metro). Before joining Metro, he was a reporter for the *Los Angeles Times* where he was part of a team that won the 2005 Pulitzer Gold Medal for Public Service for a series of articles about problems at a major public hospital. He joined Metro in 2009—the year that Metro started *The Source* to create a direct connection with its riders and keep them informed about the workings of the agency.

How does a Pulitzer Prize-winning reporter for the Los Angeles Times *make the transition to editing a blog?*

It was pretty easy. I was one of the many reporters laid off by the newspaper. At the time I was covering transportation and Metro called me about possibly blogging for them and how that might work.

At the newspaper, my job was to explain government. Actually working for the government intrigued me because I thought, generally speaking, most levels of government in the U.S. did not do a very good job of explaining themselves, which is especially true now that the Internet has come along and is so accessible by so many.

Government's inability to tell people what it does, how it spends money, and why it does what it does has had a very deleterious impact on this country. It has led to a tremendous distrust of government that, I believe, is often unjustified and has corroded the trust in public institutions that help make us into a true community.

By the same token, I happen to think that if an institution is going to take someone's money, then it owes them a good explanation of what it is doing. And if a blog can deliver that information straight to taxpayers, even better.

What kind of stories do you cover? Who generates story ideas?

Our posts are a mix. We blog a lot about Measure R projects because the sales tax increase approved by Los Angeles County voters in 2008 spurred a burst of both road and transit projects. So it's important that we explain the planning process—so

43

Figure 3.3 *The Source* provides a mix of agency news, industry updates, and transportation-related feature stories. (From: Screen capture from *The Source*. With permission.)

taxpayers can advocate for what they want—and we need to also tell people about construction impacts and those types of things. We also focus on trends in mass transit and try to answer customers' questions about the system.

As for story ideas, they come from all over the place. Some are my ideas, some come from project teams, and others from the communications department—for example, when they have a news release or issue they want out there. (Figure 3.3 shows posts from a typical edition.)

What is the mix between agency press releases and original stories?

I'd say original stories are usually 60 to 70% of the mix, although that changes day to day. The big challenge is keeping the blog conversational, interesting and, to some degree, fun.

How many stories do you typically post in a day?
> We generally try to have three to five postings a day. Not all of them are original stories. I usually do a roundup of transportation headlines based on a list published each day by the Metro library. I'll actually summarize a few stories that I think are important and offer my opinion on them.

Who writes the blog?
> Metro staff and I currently write the entire blog (I'm a contractor). We have had freelancers in the past, but we now have enough Metro staff contributing that we don't need freelancers. Using staff is also a way I try to save money.

About how many staff hours a month (your time and others') does Metro devote to The Source?
> There are a few staff members who generally give a couple of hours a week to the blog, sometimes a little more—they have other jobs first. I usually work 20 to 30 hours a week for Metro, but it can vary.

What are your goals or expectations for The Source?
> My primary goal is that people look at the blog and feel a little bit better about government—that their tax money isn't just vanishing into a bottomless well. I also want people who read the blog to feel like they're getting an honest take on things. We don't pretend that *The Source* is a media outlet; there is never any substitute for outside scrutiny. We are Metro's voice through and through, and we certainly have our own point of view on things.
>
> That said, I don't want people who read the blog regularly to come away thinking it's a tool to cannon blast propaganda across the L.A. region. We regularly publish criticism of the agency on our comments board and in our weekly wrap-ups of customer tweets about Metro.

Do you believe you have achieved these goals?
> I think we're on the right track, but I don't think we—or, for that matter, most government agencies—are all the way there.
>
> It's human nature that people don't like criticism or controversy. I saw it at the *L.A. Times,* when reporters and editors were often incredibly defensive about some stories. And I see it at Metro, where there are some—not many—who want to shy

45

away from anything that might be controversial or see every-thing through rose-colored glasses.

So, that remains one of our main tasks: persuading every-one that the best way to get people to trust the agency, use our products, and invest in it is to be straight up with the public when issues arise such as a problem with construction or fund-ing challenges.

Do you feel that you can represent Metro objectively? Can you question agency actions or policies?

I'm not sure it's my job to be objective about Metro; I'm not a journalist covering the agency. I work for the agency and it's my job to help people understand the agency's views and why and how the agency does what it does. I can—and certainly do—question agency policies and actions on a regular basis in order to understand them. I don't write everything I know just as I or any reporter doesn't write everything they know for a variety of reasons. I feel strongly that it's my job to understand what's going on so that I can answer questions about the agency and help craft a message that I think will be useful.

What kind of internal approvals do you need before posting stories?

It depends on the post. Most of them are reviewed by my boss in the communications department. Others—in particular the ones about projects—are usually reviewed by the folks working on the projects, mostly to ensure that I'm getting it right. The world of transit planning has its own peculiar language and sometimes it doesn't translate too well into plain old English!

What do you know about your readers?

I know some things, but I'd like to know more. I mostly know the numbers—how many people are reading the blog on any given day, the number of unique users, and that sort of thing. If we're going to spend money on the blog, we want to make sure people are reading it. And they are. We had more than 350,000 unique users visit the blog in 2012 and we're on pace for more than 400,000 in 2013. We've also greatly increased our presence on Twitter, which some readers prefer over the blog.

I don't know how many riders read the blog. But I do know we see something on the order of 1.4 million boardings on the average weekday on our bus and train system and I can tell that nowhere near that many are clicking on the blog each day.

The Source originally did not allow user comments. Why did you change that policy?
It was my decision at the beginning of the blog to not have comments. I had two main concerns: First, based on my experience blogging at the *L.A. Times,* the vast majority of comments would be left by 10 to 20 hardcore readers, making the comment board boring and too exclusive for my taste. I want *The Source* to appeal to casual riders who may want to leave a comment but don't want to be shouted down because they're not steeped in the minutiae of transit policy. Second, I didn't want to spend my time and taxpayer money fact checking comments as a great deal of them are filled with information that is either plain wrong or ridiculous.

I've softened my stance on this out of practicality. As per our comment policy, I refuse to approve comments that I think are egregiously wrong. In order to save time, I do let some through that I suspect are wrong or mix fact and opinion. In those cases, I hope the board polices itself. But it can be the Wild West—for sure.

In what ways does The Source complement Metro's other communications strategies?
In the olden days, Metro primarily reached consumers through traditional media—newspapers, electronic media, and even cable TV shows. The blog allows us to go straight to consumers and bypass the media, which often doesn't cover many issues concerning the agency. For example, it's always been tough getting media involved in the planning process for large projects. Planning can take years and is riddled with arcane language; actual decision points are somewhat rare. But planning is terribly important. It's when a project is shaped and molded and we want to share the studies and issues with taxpayers frequently. *The Source* allows us to do that whether or not the regular media is interested.

Does Metro use any metrics to gauge the effectiveness of The Source?
We keep an eye on the blog's metrics, including the size of the audience, to ensure that people are reading and that we're spending taxpayer money on something that people find useful. And they are. We attract new readers and get a fair number of clicks from returning visitors, too.

Figure 3.4   *The Source* frequently features transit-related photos. Here a Metro bus is crossing the Arroyo Seco Bridge in Pasadena. (Photo by Steve Hymon. Screen capture from *The Source*. With permission.)

What was your greatest success?

I think the greatest success of the blog is that we've lasted nearly three years, having published a few thousand posts and having established ourselves in the local blogosphere. I feel like a segment of the public knows more about where their money is going and we've been able to help people get around town on transit.

Any favorite posts?

My personal favorite posts are ones featuring photos of the Metro system by Metro riders. I dig photography and these photos, both individually and collectively, help inform how people see the system and the surrounding area. (Figure 3.4 shows a post from "Art of Transit," one of the regular features in *The Source*.)

What was your most popular post?

Our most popular posts were about the selection of the route for the Westside Subway Extension and the announcement of the Expo Line opening.

What was your biggest misstep?

Probably not making some decisions about what we will cover and won't cover on the blog until after we started posting it. For example, we're still wrestling with what, if anything, we should say about accidents and/or incidents involving Metro buses and trains. On the one hand, we tend to shy away from saying much of anything because the agency is a party in any incident that occurs and litigation sometimes follows.

On the other hand, it looks as if we're being evasive when something occurs—even if it's not our fault—and we don't say much about it. We take it as a case-by-case basis, which perhaps is the best route. But it means that we appear inconsistent and that's generally not a great thing.

Not every transportation agency has the opportunity to hire an editor for its blog. What advice would you give your counterparts in other organizations, especially those that may not have the same resources as Metro?

Blogs are time-consuming, so the best thing any agency can do before it takes the leap is have a good plan for division of labor. There needs to be one or two people calling the shots—where the buck ends—and there needs to be regular contributors to keep the content fresh and help build an audience.

What advice do you wish someone had shared with you?

Nothing in particular. I covered the agency before working for it and I knew that it would be like many other large organizations where I've worked—big, complicated, and at times tricky to navigate. That said, I've found the vast majority of people at Metro to be very helpful, with a particular attentiveness to routine customer issues. That's very pleasing to see.

What developments in social media do you anticipate over the next few years?

I'm not sure there will be anything dramatic. It feels to me that the creation of social media over the last decade was the big development and now we're in the stage where things will sort themselves out and the marketplace will decide what works and doesn't work for people.

I tend to think Facebook has peaked—my experience is that it has gotten harder, not easier, to use. I also think the Internet and social media will get increasingly cluttered with too many friends and followers, so it will be more difficult for anyone or any government agency to cut through the noise and reach people.

That said, I think social media is like most everything else in life. The strongest voices and those with the most to say will be heard.

MORE IS MORE: ON SITE WITH THE AKRON-CANTON AIRPORT

The Akron-Canton Regional Airport (CAK) is located in North Canton, Ohio, about 50 miles south of Cleveland. Kristie VanAuken is the senior vice president and chief marketing and communications officer and Ryan Hollingsworth is the communications and social media manager. An early adopter of social media, CAK uses Facebook, Twitter, blogging, YouTube, Flickr, Pinterest, and Google+ to engage with customers and extend its brand.

How would you say that social media has changed the way public agencies communicate with their customers, citizens, and stakeholders?

Kristie VanAuken (KVA): It really focuses on engagement. It's really foreign territory for public agencies of any type to think in the way of engagement. I feel like there are significant—whether they are artificial or real—barriers between people and the resources and infrastructure that serve them. One of our goals is to break them down.

That's one of the key differentiators for us. We think more like what you'd see in the private sector. And an airport, too, by virtue of the way that we're oriented, we generate significant operating revenue. Many of us live and die on that operating revenue. Unlike other government agencies that are very tied to public resources, whatever they might be, we're not. We have to be successful. That does create a culture that would mimic more of a private sector organization than it would a public sector organization.

With that being said, not every airport is very engaged. Then you have to get into culture, into their ties to their political structures, their management, their organizational structure. For example, we're an airport authority. And that does also lend to our flexibility and our ability to react quickly and to not ask permission from the mayor before we tweet, for example.

But you have a board of directors, don't you?

KVA: We do. And they give us great guidance but they don't try to run the airport. We're lucky that we're empowered rather than restrained by our board. They may not think in terms of engagement and delighting customers. It might be a completely foreign language to them.

Ryan Hollingsworth (RH): But it can be done.

How did you get started?

KVA: We got started when Ryan and another colleague came to me and said, hey, what about Facebook? This was in 2007 so it was really really early. We were already blogging—we were the first airport in the country to blog—we started that in 2005.

We were early adopters. And that's part of our culture, too. We want to be fast. We want to be bleeding edge, not just leading edge.

So they came to me and said what do you think about creating a Facebook group, and I said why should we? So they created a white paper to help me understand why Facebook is a good idea. The basic premise was it was mostly this is a way for us to engage with customers. And before that we were unable to engage with customers because the relationship was between the airline and the passenger. And we were cut out. We'd already done significant branding work up to that point, and this played very well into our brand, into our philosophy of engagement.

Ryan is an exceptional practitioner of social media and she was able to take our brand voice and extend it into social platforms, which was really important to our effectiveness. (Figure 3.5 shows how CAK uses Facebook to engage its customers.)

How would you describe your brand?

KVA: Very simply. Price plus experience equals a better way to go. It's that simple. Customer experience, airline prices, equals a better way to go. We feel like there's no airport in our service territory that can provide a better experience to their customers than we can.

Figure 3.5 Akron-Canton Airport uses Facebook to engage its customers. (From Akron-Canton Airport Facebook page.)

Who are you competing with?

> **KVA:** Bigger metropolitan airports. We felt like it was a very defensible and unique brand position and it turned out very much to be so. When you have that kind of simple formula it's not difficult to take that brand voice, that commitment to engagement, to experience, and let it play out in social ways or on social platforms.

Twitter seems to be a good match for many transportation agencies.

> **KVA:** I love Twitter for so many reasons. For the competitive intelligence we get from it. But also for the real-time interaction with our customers. This is a very high-level of engagement.

How are you using Pinterest?

> **KVA:** We didn't just go in there arbitrarily and start pinning stuff. We carefully thought through what is our brand, how can we make this unique to us and to our personality. How can these boards and these images really amplify and create relationships? So it's not random. We don't do anything in any

random way. We incorporate spontaneity but within the structure of our brand voice. It's very different.

Just the two of you are involved with CAK's social media activities. How much time do you devote to social media?

RH: I think the most time is spent creating the content that we're going to put on our social media. And then some days we get a lot of questions so I may spend a good portion of the day on Facebook or on Twitter answering questions.

KVA: The time varies. For Ryan, somewhere between 30 and 50 percent of her day might be spent on social. And for me it's more like 15 percent. I mostly do backup.

And then strategy. We spend a lot of time researching what others are doing, competitive intelligence, thinking about structuring our content in a way that's affirming to our brand. All of that takes time.

RH: I think we spend more time thinking about than we do actually going to our page.

KVA: We think about how we're going to get more interactions. Often people ask me, why do you care if you get more? And my response is always, more is more. More is more engagement. Why would we want less engagement? Why would we limit that?

So we set goals every year for increasing the number of interactions, increasing engagement, and increasing the baseline number, which is just likes on Facebook, for example, or number of Twitter followers.

We want those baseline metrics. But then we also want those higher levels of engagement. We don't do formal studies and go out and benchmark airports formally. But we do go out and look, and you can get a fairly good overview or impression of how we stack up against our competition in terms of engagement when you see other airports. A major international airport, for example, might have three times the number of likes that we have but maybe half the number of likes on a post.

How are you measuring goals and engagement? Are you using the built-in statistics?

KVA: We use the built-in statistics, and there are plenty in for us to get that. Honestly, what we want is to get the person to convert to being our customer, but it's hard to make that jump when you're an airport. So it comes down to the basic philosophy that we truly believe, that if a customer is engaged with us and they like us, and they feel a sense of connection. So particularly if all things are the same in a commodities business like the airlines industry. Let's say prices are the same, aircraft are the same, things are generally the same. We're going to win the business. And I don't think it's a stretch to think that. When you feel connected with a brand you're much more likely to use it than when you're disconnected or neutral even.

How many of your customers are engaged in social media?

KVA: What we know is that our core demo for our travelers matches extremely well to somebody who uses social media. The travelers tend to have more discretionary income, they're a little older, college graduates, these people are in social networks in droves.

When you started, did you have to jump through any hoops?

KVA: No. I made the decision. We're one of those places where I just said okay we're going to do this. We keep our board updated. But I'm the senior vice president here; it doesn't really need to go much higher. And particularly to somebody who wouldn't get it anyway.

But sometimes the board is who you have to convince.

KVA: Not here, though. As a matter of fact, we even pulled our very left-brained president and CEO Rick McQueen into the social forum. He does a Prez Says online forum on our Facebook page once a month.

How does that work?

KVA: People come in, they ask questions, and he answers them. That's really effective. People like that. Talk about a really effective way to break down barriers between customers and the infrastructure that serves them. You get the top dog at the organization to buy in at that level and it really creates this much flatter, more integrated, and more engaged universe for our customers that are using this place. I really like it. And he was not hard to convince. He defends it actually. At one point we were thinking about dumping it because it wasn't getting a lot of traction. And I thought, Wow a full hour just to do nothing or to answer very few questions—it just felt like a waste. So we thought maybe it isn't worth it. But when we moved it to

Facebook from our website it started getting much more popular, a lot more exposure, and we're definitely going to keep it. But when we were at that point, I said Rick, are you enjoying this and do you think we should keep this up, because it's kind of a lot of time to dedicate to something that's relatively low return. And he wasn't about to give it up.

What's your favorite Twitter success story?

KVA: We had some water get into our electrical walls and that's really dangerous. So we had to shut the power down in the building and it affected some of our service. And we used Twitter to keep the media updated about what was going on in real time. It was a very effective way for us to disseminate accurate information to customers and the media. That was a very effective case study for us.

RH: Once we started service again, the media was still tweeting that we were closed, and I was able to go in and tweet them all and say no, we're open. And they would tweet again and say they're open now.

KVA: Yeah, they're really paying attention. Mainstream media is paying attention to Twitter. More so than even going out with a big statement. We did that too, but what really moved the dial more quickly were the tweets.

Were customers the end audience for those tweets or were you sending the information to the media so they could disseminate it through their outlets?

KVA: Definitely both.

What about a Facebook success story?

KVA: The most recent is our use of social media to help launch Southwest Airlines. They just inaugurated service from CAK. Ryan came up with this idea to use a countdown, lots of posts from Superfan party, so high levels of engagement all around. Inaugural flight, all of that was put out there socially and really well received. So it was educational and delightful at the same time. (Figure 3.6 shows how the airport used Facebook to share photos from the Southwest launch party, complete with "CAKe.")

You posted a video on YouTube showing a pie-eating contest. What was that about?

KVA: : Just a little bit of fun. Why did we do it? It really plays into who we are and who Southwest is. We have very playful

Figure 3.6 When Southwest Airlines began to serve Akron-Canton Airport, marketing staff used social media to create buzz around the event. (From Akron-Canton Airport Facebook page.)

cultures. We like to have fun. I don't know if we would have done a pie-eating contest if it were a different airline.

But this was a really strong match with this airline and this airport. So it goes back to creating a different experience, to doing something that's fun, quirky, engaging. Things like that just work for us. I'm not certain that a larger airport could do a pie-eating contest and pull it off.

Even without social media each airport seems to have a personality. And you're not using social media to create a personality but to take advantage of it and enhance it.

KVA: Or to amplify the personality that already exists.

Any social media missteps?

KVA: LeBron James—he's a native son, he's from Akron. He played basketball for the Cleveland Cavaliers, he left Cleveland in that horrible decision, and about two years after the decision he was flying through here. I snapped my photo with him, and we said, "Local hero LeBron James flying through Akron-Canton Airport."

The venom that started spewing was swift and enormous. It was just fast. We actually got a lot of likes but also a lot of really, really passionate comments. People saying, "I'm never flying from that airport again! How can you say that? Don't you know what he did to us?"

So then I stepped back after the whole episode—we ended up taking the post down, by the way—if you look at where our Facebook fans are from, a good 45% of them are from Cleveland. And they didn't like that.

So I think I was quick to pull the trigger on that and not thinking as thoughtfully about out our audience as I should have.

Do you have a comment policy?

KVA: We mostly use our judgment. We're so small. It's really just the two of us. I trust Ryan to make an administrative decision about that. If she doesn't want something up there, it's not going to be there.

Do you have a social media policy?

KVA: We do. It's mostly for internal purposes. It doesn't provide us with a road map. It doesn't restrict our use of social media in any way. I go back to some corporate social media policies, which are, "Don't be stupid." That's the policy. We are just light on bureaucracy here.

When you've got good people, you trust them with your brand, and you trust them to make good decisions on behalf of your organization, you don't need a big lengthy policy saying, "Yes you can, no you can't, yes you can, no you can't."

How do you handle people who are criticizing CAK in the social space?

KVA: Occasionally they tend to have some legitimate beef about their experience here. They don't come out and say, "The airport sucks." We don't get direct hits like that. There will be an experiential issue: I lost my bags on an airline and the van didn't come to pick me up and the WiFi didn't work, and I'm mad. And so that's actually a great opportunity to engage that customer, to work it out with them, and to save them. That's our strategy.

What advice would you give your counterparts in other agencies? How would you tell other airports how to do social media right?

KVA: It goes back to something that Ryan recently shared with me that we've done intuitively but we haven't labeled. The posts need to be empowering, informative, or engaging. If they're not those things then it's probably not going to be effective. We would encourage them to use many more visual assets—photographs, video—particularly to help tell the story. We would counsel them to give their customers an inside peek at the inner doings of the airport because that's just interesting and it's someplace you can't go otherwise. So it's a really cool thing to do.

That would be the top three.

KVA: Don't be scared would be my next. Don't be scared.

A lot of agencies say something like that—don't be scared, just jump in, just do it.

KVA: It's a social world. But don't just jump in. I'm not actually a jumper-inner. I'm more of a don't be scared but don't be dumb, either. Be smart about it, think about who's the right person to create your content. Maybe a content strategy or a content plan would be a smart idea. It doesn't take weeks and months to do that. Think about who you are. What do your customers like about you? And speak to them in ways that resonate with them.

Is there anything you wish you had known when you started?

KVA: We have no regrets with anything that we have gone into. We're committed—maybe that would be another piece of advice—it's not something you can have an intern set up for you and then just let it go. You're either in or you're out.

The world is increasingly social. And I do believe that, even without a survey asking our customers.

CAPITAL CONVERSATIONS: ON SITE WITH THE DISTRICT DEPARTMENT OF TRANSPORTATION

John Lisle is the former director of communications for the District Department of Transportation (DDOT) in Washington D.C. DDOT uses multiple social media platforms including Twitter, Facebook, YouTube, Flickr, Scribd, Google+, Pinterest, and a blog to connect with constituents. The agency considers Twitter an especially effective tool for building relationships with citizens and for humanizing a city bureaucracy.

How does DDOT use social media?

The most useful platform, and the one we pour most of our resources into, is Twitter. It's just such an easy way to directly communicate with reporters and a much larger audience—the general public, people who care about transportation—and with different segments, whether residents in general, the bike community, or transit riders.

Our Twitter account really got going when we had the blizzards in 2009 and 2010. That's when we started using it aggressively and started gathering followers. We find it most useful during situations like that, and that's when we pick up the most followers and get the most engagement from the public.

What do you like about Twitter?

For the people we're engaged with on Twitter, the sharing of information is probably better than anything else that we do. We still send out press releases and participate in press conferences with the mayor and other agencies, and respond to phone calls and e-mails from the media and the public. But in terms of being able to tell people a piece of information about our transportation system that they would like to know or might need to know, there's no faster way to do it.

I compare Twitter to the other ways we correspond with people. We have a system for handling e-mails and mail that is sent to the agency. For e-mail we're supposed to respond within 48 hours, just to acknowledge we received it. But all those requests go into our system, and we are measured on whether or not we respond within 30 days. So you compare 30 days with a minute or 30 seconds, which is typically what we try to do on Twitter. It's just a very stark contrast. Twitter is much more immediate.

Do you think Twitter is letting you reach citizens that were not previously engaged with DDOT?

Certainly we're reaching people who were not engaged with our agency, at least not on this level. They were not the kind who typically would write a letter to us. The city is undergoing a transformation; there are a lot of new residents, a lot of young residents. It's a way to bring them under the tent and communicate with them.

Being responsive on social media helps chip away at the notion that we're a big bureaucracy that's not listening. People may come to the District with preconceived notions; maybe they're not expecting much from their government. We can help change their minds or influence their perceptions of government. It's easy on social media to say something negative about a government agency or a government service or the quality of the roads. But if you get a response—if all of a sudden you post something about a road and your local department of transportation chimes in and says, "Where is that? How can we help you?" then that's a real and quick response. It's the same thing that businesses are using to engage customers and mitigate negative perceptions or feelings about things. So it's very useful in that respect.

I think that a lot of the people we're engaging with through social media might not be engaging with us if we weren't active. Those conversations would still be happening about us, but we just wouldn't be part of them.

How does DDOT compare with other city agencies?

We were one of the first in the DC government to have a really robust Twitter account and social media presence, and we probably have more accounts than most if not all other DC government agencies—for better or for worse.

Some just push information out and have less engagement. Some don't follow anybody. There are different approaches. What we've tried to do is use Twitter more like people use Twitter. Which means having conversations with people and trying to respond to every question, if we can. Engaging with whoever wants to engage with us as best that we can. We can't make everybody happy but we can certainly communicate with them.

How much time does it take?

After a hurricane or a major event, it consumes a large part of the day. The day after Hurricane Sandy, I was certainly doing other things, but I would say from the time I got up at 6:00 in the morning until the time I went to bed at about 10:30, I was monitoring Twitter, responding to people, forwarding questions or service requests. That sort of thing. And that's not unusual for an event like that.

I drove around the city and was tweeting photos from some locations. It takes a lot of time in a storm like that. One of the reasons that's a hard question to answer is sometimes when you're working with tools like TweetDeck—it's like Outlook for

Figure 3.7 Potholepalooza is DDOT's annual campaign to identify and repair potholes. (From Potholepalooza Facebook page.)

> e-mail—it's open on your desktop and every time a tweet comes in you see it. So it's just a regular part of the day and it's mixed in with other duties as assigned.

How do you handle service requests for issues like downed trees via Twitter?

> Street trees are the responsibility of our urban forestry administration, which is part of DDOT. One of the challenges of social media, is that we do take service requests sometimes via Twitter or Facebook. We've done it for our month-long Potholepalooza campaign, where we ask the public to help us find pothole locations. And one of the ways they can do that is by tweeting us the location. And then we put that in our system and the crews go out and fill it. (Figure 3.7 shows the Potholepalooza Facebook page.)
>
> With trees, we've also done some of that. But there's some pushback internally when people see an e-mail coming from you and it says "Tweet forwarded." And they're like "Oh no." We have a system for tracking service requests, so you have to be cautious about not going outside of that. So we try to push people towards the 311 system for reporting trees down. We do try to discourage citizens in certain instances from tweeting us

a service request. We'll take it and we'll forward it on but also ask, "Can you call this into 311?" That's how our tree managers track all the requests. You still need to get the information into the system somehow. In some cases, a request will come in through Twitter and we'll sit down at the computer and enter it into the system and then that's done.

We might get a request for a sign down at this intersection and we'll forward that onto the sign team but you have to be cognizant of the fact that these requests need to be entered into the system and tracked like everything else.

How do you measure the impacts of your social media activities? Do you track effectiveness?

No we don't.

First of all, I know we have a good reputation for our social media efforts. We have a good following and the sentiment is positive. So I'm not worried about whether we're being successful, because I think we are. But the second piece of that is, it would be great to know exactly what our influence is, what our reach is, how effective we are, what worked, what didn't work. All the stuff you do with analytics would be great. We have looked into it, but we haven't pulled the trigger and spent any money on doing that.

The main reason why it would be useful to have all that stuff is that a lot of what we do is probably still hard to explain to management and to others within the agency. For instance, we could probably use a social media manager. So if you want to make the justification for a position like that, or if you want your boss and others in top management to understand how much time and how beneficial it is to do social media, it helps to have some measures and some analytics to back you up.

The storm and the earthquake a year ago are good examples. I saw how the school system used social media to communicate with parents during the earthquake. They had good measures of their reach during the storm or after the hurricane. It would be helpful for internal reasons to have some of those measures, but we're not a for-profit agency, we're not a business, so I don't think we need to know all the ins and outs of whom we're reaching and how we're reaching them. But it would be helpful to know, and I certainly would like to know what tweets worked and what didn't.

I'm going to look into some of the free tools out there to see what I can measure from the last few days dealing with Sandy. It's interesting. We gain followers slowly on a normal basis but during the storm we were adding them at a pretty good clip.

Do you have to do any kind of archiving or record keeping?

We do not. Maybe at some point policy and regulations will catch up with social media use but, no, we have been for the most part left alone.

Any institutional concerns?

If you're going to be engaged, and you're going to be communicating with the public with these tools, which are so immediate, you're going to mess up every now and then. You have to be careful. You can't be insulting; you can't be rude to people. But occasionally you might tweet something that in hindsight was inappropriate or maybe you shouldn't have done it or maybe you should have run it by someone else first. So that stuff's going to happen.

There's one posture you can take, which is we're going to eliminate any chance that something like that could happen. You have to get six people to sign off on your tweet before it goes live. Or you can accept that there's an inherent risk but the value outweighs that. And that's our posture, so far anyway.

How do you handle people who attack DDOT?

We try to engage them as best we can. My philosophy is, let's see if we can turn this around. If someone says, "I can't get through to anyone at DDOT," I'll give them my e-mail address or phone number so they can call me and I'll try to take care of it. You can't always solve their problems; sometimes you're only as good as the services the agency provides. Say somebody asks us for a sidewalk repair. Well we have a backlog of those; we're not going to be able to do that. Maybe we can make sure it's on the list and give them some information about how long it's going to take. But they're not going to walk away with a sidewalk.

So you're limited. And also if somebody else drops the ball and you get a service request for something and you pass it along and that department doesn't do anything, the agency's

Figure 3.8 DDOT uses Twitter to answer citizen questions. Here John Lisle invites a Twitter user to follow up his question in more detail via e-mail. (From DDOT Twitter account (@DDOTDC).)

going to look bad and there's nothing you can do about it. But if there is some way that we can turn around that perception, we try to do that. And I have found that people—as negative as they may sound at first—the fact that they got a response from their government, that somebody was paying attention, makes a difference. They might say, "Hey, thanks for responding, wish we could get this done, but I understand that it's out of your hands." So I think that it is possible to at least improve the way they're thinking about the agency by engaging them. (Figure 3.8 shows a Twitter conversation between DDOT and a citizen with a complaint.)

Any advice you would pass on to other DOTs?

As I said, one of the things I like to do is use social media like people use it—share links, try to be funny sometimes, certainly a treacherous area to get into but we certainly have tried to be engaging with people. We're occasionally accused of being snarky, not being bureaucratic, but we engage with people as much as we can, as part of their circle on Twitter.

You can only do so much. The things like GooglePlus and Pinterest and tools like that—it looks good that we're on there—but there's limited value for some of that stuff and you can really get spread thin. You figure that out as you go along. Is it worth my time to be pinning a picture on Pinterest? Maybe yes, Maybe not

The other lesson is, think twice before you tweet things. Count to ten.

FROM ZERO TO NINETY: ON SITE WITH THE CENTRAL OHIO TRANSIT AUTHORITY

Jamison Pack is the director of marketing for the Central Ohio Transit Authority (COTA) in Columbus. COTA connects with its riders via Facebook, Twitter, and YouTube and is beginning to incorporate Foursquare. The agency also has Facebook accounts for the Ohio State University population.

COTA recognized the importance of social media when its bus operators staged a strike around the Fourth of July holiday in 2012. The agency used Facebook to keep its riders informed during the work stoppage and this helped create a community of engaged followers. Just a week later, a freight train derailed adjacent to COTA property early in the morning and interrupted service for hours. But instead of communicating with their new online community, officials were caught off-guard and did not update COTA's social media pages in time for the morning commute. Jamison Pack talks about what COTA learned from these incidents.

Transportation agencies now use social media for all sorts of reasons. How did COTA get started in the social space?

With caution. COTA formally engaged in social media in spring 2011. However, once it became apparent that our represented workforce might exercise their right to strike, we began working on our communications contingency plan. Once the union gave its 10 days' notice of intent to strike, we knew that there would be many concerned stakeholders (taxpayers, customers, and employees) seeking accurate information regarding our service and the status of the strike. COTA created talking points to use with our stakeholders and enlisted the services of a social media consultant who modified these points to use on Facebook and Twitter. It's one of the best decisions that we could have made, because as soon as the union announced the strike we started to see activity spike on social media.

For the first time, we had the opportunity to engage directly with our audience in a way that we couldn't do with traditional media. We were able to educate and inform; clarify false information; provide critical customer service and participate in the conversation. We had a duty to help our customers prepare for no service, but in doing so, we also had a duty to let our taxpayers know what we had done to negotiate in good faith in order to prevent the strike. Our goal was to use the 10 days before the strike as a final push to prevent the strike.

So the 10 days was the time between the notice of the strike and the start of the strike?

Yes, they gave us their notification on Friday, June 22. That meant the strike would go in effect on Monday, July 2, at 3 a.m.

So we had four statements that we wanted everyone to know about. First, that we were responsible stewards of taxpayer dollars. Because we rely on the sales tax, we needed to let folks know exactly how we are managing public funding. We wanted people to know that we respect our employees and we treat associates fairly. And we wanted people to know that we had been bargaining in good faith. Finally, we wanted people to know that we remained at the table and we wanted to continue the discussion. However, we also have a duty to our riders to let them know that on July 2 they had to make alternative plans because only the union can decide to call off the strike.

How did social media factor in?

Our stakeholders went to social media immediately to engage with us and other pages (i.e., local TV media and elected officials). Throughout all the events, we monitored our pages and other pages, like the TV stations. It became one of the best channels to keep in touch with customers. As we monitored, our social media consultant would text me updates and posts made by other people. And if it fell within one of the four talking points or was a specific service-related question, we responded. Also, the social media community would engage and in many instances, appeared to understand and support our position. You could see the dialogue taking place, where we pop up and where we let the community continue with their conversation.

The depth of the interaction was significant; some posts had 20 or more responses from a wide variety of people.

Our activity on Facebook and Twitter the weekend before the intended day to strike was very active and much more emotional. The union agreed to meet on Sunday July 1, at 1:30 p.m. and customers were waiting by their computers to know whether or not there would be a strike. Unfortunately, that meeting concluded on Monday, July 2, at 1:30 a.m. A tentative agreement was signed but the union decided to strike. We knew it was a possibility; however we hoped that it wouldn't happen. At that point, we needed to let folks know that the union was striking and people needed to make alternative plans.

We also decided to make a clear shift in our communication to be transparent about the details of the contract and to do our best to set expectations with our customers. Because we believe in our represented workforce and we support our represented workforce, we wanted to demonstrate what exactly the package had and you can see these details in our posts. The public was already talking, but now the public was really responding passionately in support of ending the strike as soon as possible.

The community went beyond our Facebook and Twitter accounts to our Mayor's and TV stations. They were voicing their frustration and asking for help; lobbying for the strike to end as soon as possible. After the union membership voted down the tentative agreement on the afternoon of Monday, July 2, union members were still striking and picketing.

We offer a major service on July 3 for the fireworks in Downtown Columbus. This was the first time in 25 years we did not offer it. There were a lot of frustrated customers that use our service just for this one event every year. It created another wave of social media communication. And then on the Fourth of July, the union decided to revote, approving the contract and returned to work on July 5, 2012.

So, as you can see on our Facebook page, we let folks know. On the fourth, we were posting updates and responding to customers. Our board was convening and we even posted a video of the statement from our board chair from the meeting to let folks know the strike was over and that it's time to mend fences and get back to work.

67

Figure 3.9   COTA's Facebook page shows agency posts about the operator strike juxtaposed with service updates after a train derailment less than a week later. (Screen capture from COTA's Facebook page.)

> And then, exactly one week later, a train derailed partly on our property. We were not able to offer our full service and we did not do a good job of communicating. We were not on social media; there was nothing on our website. During the strike we quadrupled our Facebook audience and doubled our Twitter followers; we had built up this audience of people who were used to having lengthy conversations in our space and they just expected to receive information from us and they got nothing. (Figure 3.9 shows COTA's Facebook page in early July.)

So what happened?

> We weren't prepared. The train had derailed in the middle of the night and our service was significantly compromised the following day. So many people were focused on the evacuation of the area (evacuating our employees as well as surrounding residents) that no one thought to communicate with our team to get information online.

Because our social media efforts are managed by the marketing department, our public relations team focused their efforts on responding to overwhelming requests from traditional media. Meanwhile, customers were hitting our Facebook, Twitter and website wondering what was going on and if their bus was going to arrive or not.

By the time I knew what was going on it was too late to communicate with customers seeking information for their morning commute; we were already behind the eight ball. The other factor hurting us was that we were no longer engaging with the social media consultant; it was back to business as usual. However, the problem was that we built a significant audience leading up to and during the strike; they expected us to be there during the next crisis. We tried as best as possible to put accurate information on Facebook and Twitter, but we also apologized and committed to doing a better job communicating with our customers. That's one thing that social media is a great tool for; it allows you to publicly own your mistakes. Most people are reasonable. They know that we can't be perfect all the time. We had to own this one. We just said that we hear your frustrations and apologize and we want to try to prevent this from happening again.

So now we're working on making our social media communication better.

Here's a supportive note from someone on your Facebook page: "Hey COTA— Didn't get a chance to thank you for the quick response during the derailment yesterday."

What I like about that comment is other people see it too. And maybe it helps people become more understanding.

Both Facebook and Twitter let me post as COTA or as an individual. Every once in a while I prefer to post as a person. Sometimes I see people overly critical of our service—and they may have every right to rant—but they're not being constructive about it. And I'll contact them from my account and say, "Hey, I saw this and I'd like to help you with this." I get one of two reactions nine times out of ten. I get no response at all or I get "Oh my gosh, I can't believe you contacted me. Do you know how good it feels to know you're really listening?"

I love being able to do that and let folks know—"You know what? We actually have a process for your complaint and I'm more than happy to move it to the right department." And then explain to them in order to do that I need a little more information.

So within one week you had your greatest success and your biggest misstep.

Yes, we did.

What would you tell people in other transportation organizations?

I think it was really surprising for this organization to realize how powerful a tool social media is. We surprised customers by engaging in tough conversations and we even surprised our own staff. It was an opportunity to own our mistakes. It was an opportunity for us to listen to our customers and personally respond.

I think the biggest thing for me has been the ability to demonstrate the power of social media in order to get that internal buy-in. As marketers we know that we have an opportunity to build better relationships with our customers; we have an opportunity to boost loyalty. We have an opportunity to build a community of supporters both online and offline. Plus social media costs significantly less money compared to traditional marketing tactics but the value is huge.

And social media gives us the ability to influence people. The private sector can influence sales. In the public sector we have the ability to influence someone's decision to get on the bus or take their car. And it's our job to get more people on the bus. If you know anything about Columbus, you know this is a very car-friendly place. We don't have the congestion challenges that other communities face. So it is very difficult to convince someone that transit is convenient. We have an ability to influence their opinion online, and that's powerful. So the strike and the derailment taught us that we have an opportunity to be smart in the social media space.

What was the biggest institutional barrier you faced at COTA?

We don't deliver well online yet. Our website is flawed and up until the strike I didn't have much support for being on social media channels. I had their OK to get started but people didn't really understand the value internally. And I knew

70

that—it takes a long time. It's a cultural shift. I've honestly just taken the position of being patient on it.

When we came to the strike, we really had no choice. We had to leverage Facebook and Twitter or let someone else who didn't know the details fill the space with chatter. We had to go from zero to 90 on the spot. And so yes, it's not the typical approach that people take. It's certainly not the ideal approach. You know, you get your ducks in a row and you make sure your promotion is across all platforms. We didn't have any of that. We also didn't have the luxury of time to get our plan in place. But now that we have created customer expectation to be available through social media, I have internal support. I have had people on our leadership team indicate that they agree a significant reason we were able to influence public opinion was because of being on social media and our response to comments. And now we are proceeding with creating a social media plan that will be hope-fully comprehensive in our approach and it will consider our offline promotions as well as our online promotions.

Other advice?

Don't wait until you have a crisis to figure out what the heck you're going to do. I was lucky that I was able to engage with a social media consultant, and it was helpful that I had experience in the social media realm before coming to COTA. But without that, a couple of things could have happened. We could have just shut down our Facebook page. And what kind of message does that send to people? These days, not being on social media is the equivalent of traditional media's "no comment." Or we could have potentially had a whole bunch of angry customers ranting. It had to be done.

Are you saying that the conversations will go on with you or without you?

That is very true. You can either choose to participate and show that you're listening or you can not engage at all. Not engaging comes with negative consequences. It gives the perception that you don't care and that you're not will-ing to do anything about it. It also is a missed opportunity to clarify inaccurate information and demonstrate your posi-tion. Equally important, it's an opportunity to acknowledge people's experiences with us. People have a right to be frus-trated. It's OK if we acknowledge that we're not perfect. This

71

Figure 3.10 Example of COTA's use of Twitter to stay in touch with riders about travel conditions. (From COTA's Twitter account (@COTABus).)

is a complicated time for public agencies. The level of accountability and transparency is critical to the continued existence of public agencies. And one of the best ways to demonstrate that you are accountable and responsible and transparent is to do it on social media. (Figure 3.10 shows how COTA uses Twitter to engage with its riders.)

Who creates COTA's social media posts?

It's an intern and me at the moment. However, we are re-engaging with the social media consultant and created a new position, online marketing manager, who will be responsible for all of our online properties.

What's next for COTA's social media strategy?

We are in the process of claiming our bus lines and shelters and bus stops and facilities on Foursquare. We have 4,000 bus stops so we are identifying the lines and stop locations that our customers are using and checking in most frequently. We're going to where people already are and have already identified our service. We are renaming these places and claiming them so that we can own that property and real estate and still allow people to engage with us. As we enter our Short Range Transit Plan, we're using these Foursquare locations to encourage customers to take our online survey. These days it's harder and harder to get people to come to public meetings. We're eager to see the results of the Foursquare tips function.

Do you plan any Foursquare promotions?

We would like to engage people in some fun activities. We already experience people leaving tips and providing what is mostly positive and helpful information. From a marketing perspective I like that people want to be associated with us. It's a nice word-of-mouth tool for us that doesn't come from our mouth. It also helps build loyalty. Even though it's a tiny impression it helps create a deeper relationship.

Any final thoughts?

I think there are many reasons why this experience could be helpful to other agencies. A lot of us go through contract negotiations on a regular basis. It is a very delicate balance of engaging your public, telling your story, and engaging your employees. One of the biggest struggles around the strike was demonstrating that we bargained in good faith and that we support our employees. And then once it was all done, being able to reunify, move on, work and operate ideally from one voice. It is a delicate balance. I hope that other agencies are thinking about it; I know that communicators do, but they have to get buy-in from their CEOs and other people who make decisions to be successful.

NOT GOVERNMENT AS USUAL: ON SITE WITH THE ARIZONA DEPARTMENT OF TRANSPORTATION

Nicole Sherbert and Timothy Tait are assistant communication directors at the Arizona Department of Transportation (ADOT). Tim runs the Office of Public Information and oversees ADOT's Twitter feed. Nicole manages the Office of Creative Services, which is responsible for ADOT's other social media activities. These currently include a blog, Facebook page, YouTube channel, and accounts on Flickr and Pinterest. As a statewide agency, ADOT uses a combination of active engagement and relevant content to keep citizens informed about travel conditions throughout Arizona.

When did ADOT start using Twitter?

    Tim Tait (TT): Part of what I manage is our outbound communication from our traffic operations center. That's all of the incidents that happen on highways—crashes, vehicle fires, weather events. We were looking for another avenue to push all that data to drivers.

    So, in 2008, I started the Twitter account and it took off pretty quickly because of the very regular postings we made on traffic incidents. It wasn't every crash but it was crashes that had an impact on traffic flow. And in a big state like Arizona, and in a big metro area like Phoenix, that's quite a few incidents every day.

    So Twitter was an ideal resource to move that information and to get it to drivers. Its text message capability was well suited for driver information. Of course we have the disclaimer that tweeting and driving don't mix; that's a concern that we have. So we try to mitigate that as much as we can.

Are these messages automated?

    TT: We automate nothing. Everything has human interaction. It's edited by a person, it's inputted by a person, and it's sent. I do not believe in automated messages. I think they're dry and they totally miss the whole point, which is engagement.

Who crafts the messages?

    TT: I was the solo tweeter for quite a while, but now we have several people who have access to Twitter including duty officers who are stationed in the traffic operations center.

So there are people in the traffic operations center who find out about an incident and immediately disseminate the information?

    TT: Yes. It's much more immediate and it's much more engaging because they have the time to put a personal touch on the messages. They're doing a lot more interaction directly with drivers, providing detour routes. They can also share screen grabs from freeway cameras. We can tell people about congestion, but now we're showing them pictures of what the congestion or what the crash looks like..

Do you have just one Twitter account or do you have multiple accounts for different regions?

TT: We have maintained one primary Twitter account for all of our incidents and other communication. The philosophy has been not to splinter the audience. We worry having multiple accounts really fragments the audience and we lose out the engagement opportunities that we otherwise would have. Sometimes we lose followers because it's just too much information for them, but our track record shows that the philosophy is serving us pretty well. One of our big emphases is maintaining that audience so when we've got something important to tell them we can tap into them.

Nicole Sherbert (NS): We've extended that same philosophy to all of our social media. We're one agency and we are statewide. Part of my job in the creative services office is to brand the entire agency and introduce people to everything that we do. We plan the roads, we build the roads, we maintain the roads, we issue the licenses to drive on those roads and the permits for trucks to drive on those roads. Social media users may think they only want traffic information but we are extending to them the information that we want them to receive.

Did you have any challenges getting social media up and running within the agency?

TT: When I started I didn't really ask for permission. I just did it. And nobody told me to stop. Starting the YouTube and Twitter channels was a natural progression in the overall strategic communication plan for the agency. These were not one-off efforts; these were part of a bigger effort to move our communications efforts forward.

Did you encounter any institutional issues?

TT: The only institutional issue that we had, and still have, is that our own employees don't have access to read these social media sites at work. It requires high levels of approval to get access to these sites while at work. So Nicole may have a great blog and a great Facebook page full of resources that would be informational for our employees to help turn them into ambassadors for the agency, but they can't get to those pages. There's probably lots of good solid reasons for that in the IT world but from a communication perspective that's something we continue to struggle with a little bit. How do we reach our employees with these tools without trying to recreate something or putting something in a walled garden? We've got a handful of people who are working hard to generate the content for these outlets and then the 4,500 employees that we have in the agency can't benefit from it unless they seek it out on their own time.

Do you reply to comments from citizens on your social media accounts?

TT: Absolutely. We go looking for comments. We reply to comments that are sent to us, and I have search parameters set up so we can find comments. Maybe someone didn't have a good experience at a Motor Vehicle Division office; we want to find out what that was, what we can do better, and if there's a way we can help them with that transaction. We reach out to people and there are occasions when I'll send somebody my e-mail address and say, "Hey, send me an e-mail. Tell me the whole story. Let me help you out with this." So we not only respond to messages on Twitter but we go looking for people to engage.

NS: We have a very interactive Facebook account. We've done a few things to try to increase our engagement. We started doing a photo quiz called "Where in AZ?" Our video team will take some obscure roadway picture and then we post a "Where in AZ?" And we average 25-30 responses on those, which is great, and we get a lot of people who share and like them. (See Figure 3.11.)

We also respond to every comment. If it's negative, we'll try and address it and at least thank them for their comment. And overwhelmingly we have more positive comments than negative.

Can you give me an example?

NS: Like most of the country, Arizona has been in a financial downturn. We had somebody who posted on Facebook and, in a very negative tone, said "I thought this state was broke. Why are we putting up new signs on I-17? Way to go, ADOT." ADOT replaces those signs for safety reasons, and the project was funded through a grant for safety enhancements. So we took all that information, wrote a blog post about it, and then replied to the comment on Facebook. We said we thought it was such a great question that we went ahead and wrote a blog post explaining why people are seeing those new signs on I-17.

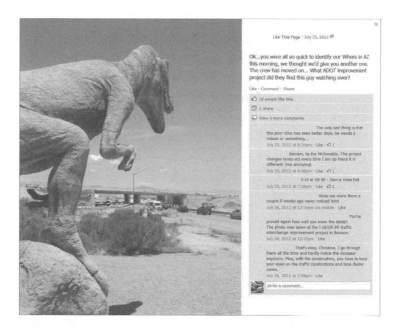

Figure 3.11 Arizona DOT entertains and engages citizens with features like this photo quiz. (Screen capture from Arizona DOT Facebook page.)

The guy thanked us profusely and is now one of our biggest fans. He shares our important posts, he comments positively, and he responds on the "Where in AZ?" So we've taken what was a negative comment and really educated the individual and turned him into an advocate for us. So he is now one of the people who are engaging with us and helping us to spread our message.

I understand that a haboob is an intense storm that creates a wall of dust or sand. Tell me about haboob haiku.

NS: Tim spent months working with the Department of Public Safety and the National Weather Service to see how we could improve our messaging around dust storms. They're called haboob because they're a different kind of dust storm. They build a wall of dirt that can be a mile high; it will come out of almost nowhere and turn visibility to zero in seconds.

Last year we had a huge crash on I-10 with one fatality. So when Tim was working on the new messaging, they changed the message to: *Pull Aside—Stay Alive*. The message in the past had always been what to do once you find yourself in a dust storm. And we were changing the messaging to say don't drive into it in the first place.

So while Tim worked on the messaging, my team came up with the creative side of the campaign. We had a bunch of different ideas from the creative, and it just popped out one day: "Haboob Haiku!"

We launched the entire public safety campaign on Monsoon Safety Awareness Week and we planned a five-day effort to push messaging. On the first day we released the Pull Aside—Stay Alive public service announcement.

The second day we announced the haboob haiku through the blog and through a press release. And it really picked up. We got quite a few that first day. We'll be the first to admit we were populating it ourselves at the beginning; so we started it organically and it just took off. On Day 2 the Associated Press called. The next day it was Reuters. By the time it was said and done over 500 stories had run in daily newspapers. CNN, BBC, The Weather Channel, all had done unique pieces about haboob haiku.

TT: It really took off and it absolutely surprised us. Part of it was because some really clever entries were being submitted. But we were also getting a lot of acclaim for the fact that it was a clever approach. It wasn't government as usual. It was something different. And I think they really appreciated that

NS: The most important take-away was that it was fun, it was original. It got people talking, it got people laughing, it got people paying attention to what we were doing. But most importantly it had people writing and talking about what to do in a dust storm. And that was the first time we really saw engagement of people taking our messaging and repurposing our messaging for us and spreading it to an international audience. Haboob haiku was a tool where we used our audience to help spread our message.

**This Winning Haboob Haiku was Selected
from More than 500 Entries**

Dust blows, swirls and grows
Roadways become danger zones
Pull over, lights off

@Lisa4FLA

Some agencies get a big bump in their social media followers after a major event like a snowstorm or an accident. Did you have anything like that?

TT: Funny you should ask. This past March we had a significant snowstorm that caused the closure of Interstate 40 and Interstate 17, which are the major corridors in Arizona. The storm crashed the 511 system, which is our traveler information center, so the only way to communicate with drivers on the road was through social media. This was over a weekend, and I was at my computer at home for 16 hours just responding to travel inquiries, telling people what roads were closed, giving them alternate routes, telling them how long the road was going to be closed. It goes back to engagement on that individual basis. But for emergency communication that's a great example of what we did to get the job done. And it's because our followers were relying on that.

NS: And while Tim spent Sunday at his house on Twitter, I spent the day at my house on Facebook. And we had similar situations. People shared things constantly. We gained about 200 followers on Facebook that day. And it was an effort. Tim and I were both working about 12 hours during the day and up at three o'clock in the morning checking to see if roads were reopened because social media was the only communication that was available.

Any other examples?

NS: We had a tanker collision on I-10 just south of downtown Phoenix during morning commute. The tanker caught fire and the gas burnt for hours. After the fire was put out, we saw that part of the road surface was damaged. Crews worked overnight, repaving and restriping the road, and did a great job to open

79

Figure 3.12 When a tanker crash shut down Interstate 10, ADOT staff used Twitter to keep the public informed about changing road conditions. (From ADOT via Flickr, licensed under Creative Commons Attribution-NonCommercial-NoDerivs 2.0 Generic [CC BY-NC-ND 2.0]. With permission.)

the freeway the next morning, but I-10 was closed down for an entire day. I was Facebooking, Tim was tweeting, and we were making updates all day long. We saw a significant increase in our followers that day. (See Figure 3.12.)

TT: We built a lot of good will that day. Because we were very open and transparent in our communication and it was free-flow communication. It wasn't just two updates during the event. It was regular updates, and I think that built an incredible amount of good will. And we ended up being the heroes on that day for getting it done so quickly and efficiently.

NS: But it was because we were explaining what happened. Because you would think, "OK, the tanker's gone. Why is the road still closed 20 hours later?" And it was because we had to completely reconstruct the road where the surface was damaged.

Some agencies adopt a snarky tone for their social media posts. ADOT's are conversational but they don't have an edge.

TT: We've really tried to stay away from that. We're much more customer-service oriented in our tone. We don't want to alienate anybody or to have anything read out of context. We're light-hearted but we try to be very careful with our tone. We have over 17,000 followers now on Twitter so we try to be very careful with how we're approaching those folks.

What sets ADOT apart?

NS: First, why do we use social media? And we do it for a lot of reasons. We do it to educate people, to grow our audience. But I think expanding our audience is a huge element of my portion of it, which is where the blog, YouTube, and Facebook come in. We have major events on our freeway system. Last year we closed down I-10 for 10 hours for a bridge removal and the only way around was a 67-mile detour. In the past we would have to rely 100% on the media to carry that message for us. So our goal is to grow our audience, so that when we have a major event, people who are listening to us and already paying attention can help us spread that message. They're not only getting the word themselves but we ask that they share it with their social networks. So we can exponentially expand the amount of people we're reaching.

That said, media of course is still king, but the media is following us on social media as well. So when they see an important post, they can share it with their audiences, again, exponentially increasing our reach. It's great for the planned closures we know about and want to mitigate the impacts of, and great for unexpected events whether that is a tanker collision that's closing down I-10 or whether it's a dust storm.

Other success stories you want to make sure you highlight?

TT: The big winter storm this past March and the tanker fire, those are two really good examples of how we've used social media successfully, and how we picked up followers, and used it as a tool with the media and really demonstrated internally how it's a tool worth spending time and energy nurturing.

Any missteps?

TT: Once and only once I posted a personal message on the ADOT account. I only made that mistake once, and then I set

81

up my own safeguards. But that was pretty early on and thankfully it was a pretty innocuous post and I caught it right away and was able to explain it away. That's probably been our biggest misstep and that was a learning curve on my part.

Any advice to other DOTs?

NS: To be successful, you have to do two things: You have to be social, which is the engagement element that Tim was speaking to, but you also have to have great content. You have to have content that resonates with people. Pull Aside—Stay Alive was a great message because it had that "Pull aside" in there. It directly told you what to do. It's something that is so on trend right now here in Arizona because our dust storms have been so severe in the last several years. So we had a message that resonated with people but behind that message we had great creative. The video that we used—which has had several thousand views on YouTube—was an excellent video. It was very well produced; it was very compelling. So you have to have great content. You can be social all you want, but if you don't have good messaging and good content, no one's going to pay attention. At the same time, you can have great content and great creative and great messaging, but if you're not engaging people then they're not going to listen to you.

So those are my two biggies: content and engagement.

TT: The best piece of advice we can give is don't be afraid to engage. Keep the social in social media. By ignoring those comments and that opportunity to engage directly with followers, the organization really misses out on a prime opportunity to create ambassadors for the organization. And it doesn't require a whole lot of effort. It doesn't require a whole lot of risk. They're already ticked off at you; sometimes just acknowledging that can help to nurture a relationship.

You're not going to convert everybody, but engaging at least shows that you're paying attention. And in this day and age that's an important theme for government—to let people know that you're paying attention. You're listening; you hear what they say. And hopefully what an agency hears informs what it does. We try to make sure that our followers realize that what they have to say to us matters and it can make a difference.

REFERENCES

ADOT via Flickr, (http://www.flickr.com/photos/arizonadot/6332523372/in/set-72157628098466374. With permission.)

Akron-Canton Airport Facebook page. http://www.facebook.com/photo.php?fbid=10151273321931033&set=a.87945796032.79920.84507856032&type=1)

Arizona DOT Facebook page. http://www.facebook.com/AZDOT. http://www.facebook.com/photo.php?fbid=363411793730530&set=a.137665612971817.29041.117553941649651&type=1&theater)

COTA's Facebook page. http://www.facebook.com/cotabus)

COTA's Twitter account (@COTABus). https://twitter.com/COTABus/status/284026322987130880)

DDOT Twitter account (@DDOTDC). https://twitter.com/DDOTDC/status/299195425385218051)

Potholepalooza Facebook page. (http://www.facebook.com/Potholepalooza)

Southwest Airlines Akron-Canton Airport Facebook page http://www.facebook.com/akroncantonairport. http://www.facebook.com/photo.php?fbid=10151028759421033&set=a.87945796032.79920.84507856032&type=1)

The Source http://theSource.metro.net/2013/02/05/turnstile-will-be-latched-tomorrow-at-wilshirenormandie-station/

The Source http://theSource.metro.net/2012/12/27/the-art-of-transit-361/

Washington State Department of Transportation via Flickr. http://www.flickr.com/photos/wsdot/8125807832/

WSDOT Twitter account @wsdot. https://twitter.com/wsdot/status/248865723450003457)

4

Using Social Media to Connect with Customers and Community

Social media is all about connections. For transportation, social media often takes the form of connecting with citizens, community stakeholders, members of the media (traditional and new), and public officials. This chapter focuses on four of the many ways that transportation agencies are using social media to forge those connections.

In the first section, Matt Raymond and Robin O'Hara describe what they call the Four Es of social media marketing: enticement, exchange, engagement, and experience. They draw upon examples from transit operators, airports, and departments of transportation to show the value of social media as a marketing tool.

Community activists Ashley Robbins and Andrew Austin write about their experiences using social media to help build grassroots support for transit initiatives in Atlanta and Seattle, respectively.

Social media can be an effective strategy for obtaining community feedback during long-range planning efforts. Jody Litvak details the experience of incorporating social media into the outreach process for the Westside Subway Extension in Los Angeles and Jennifer Evans-Cowley summarizes the role of social media in developing a long-range transportation plan in Austin.

Finally, Susan Bregman addresses the use of social media in providing real-time and emergency communications and Ned Racine recalls his experience using social media to help inform and educate Los Angeles residents about a major highway construction project known colloquially as Carmageddon.

COMMUNICATING AT THE SPEED OF LIGHT: USING SOCIAL MEDIA AS A MARKETING TOOL
Matt Raymond and Robin O'Hara

Yesterday's customers relied on printed newspapers for information and were influenced by advertisements they saw in publications and on billboards. They viewed local news shows on network television or listened to the radio to assess traffic conditions. They used printed bus and rail schedules or called a live operator to gather necessary information. Yesterday's customers relied on signs, maps, and posted directions to determine their current locations and where they were going. They paid for the transit system by cash or by token, often not sure about the cost of the fare or whether they were on the correct vehicle until they asked an operator. These historical forms of accessing information are slowly making way for a more contemporary, viral, and powerful type of marketing.

Today's customers rely on the web for information that they access through their computers, tablets, smartphones, and other mobile devices. Their friends in online communities and the people they follow on social networks influence them. They use quick response (QR) codes in printed materials to go directly to a company's website where they can connect with others through blogs or message boards. They skip traditional commercials and access social networks even on their televisions through digital recording devices where they can interact with live shows and post their opinions and comments. The radio stations they listen to are preprogrammed, downloadable, and also offer interactive opportunities. Today's customers use the web to get automated trip planning information. They may enjoy a live online chat with a customer service representative from their desktops, laptops, or tablets.

Today's customers get service alerts and road construction updates via texts or tweets. They use their smartphones to obtain schedules and vehicle location information such as when their next bus is coming—in real time—and can post or tweet the information to their personal networks in a variety of ways. They get step-by-step directions for bus, auto, or foot transportation and can send the directions to friends with the push of a button on Facebook.

Digital signs are increasingly available and offer instant networking access through touch screens. Consumers find and pay for parking with mobile apps and, with a minute's download, area maps become interactive and link to local restaurants or area attractions where a list of customer

86

reviews may be accessed. The transit pass is now a smart card with a digital memory chip onto which customers may add value online using a credit card or PayPal account. The back of the card lists a website where customers may go to post tips, ask for help, or talk about their traveler experiences using the card.

Accurately marketing to today's customer is even more important because every transaction is tracked and customer experiences may be digitally photographed or recorded and then blogged about, posted, or tweeted. For this reason, this chapter begins with a section describing how transportation agencies use social media to market to today's customers.

A New Social Paradigm

Social media has drastically altered how transportation services are marketed today. Yesterday's model required extensive resources and long lead times and had few outlets through which to communicate. Today's model is a complex web of consumer technology that customers interact with and access in real time. Influence and information happen instantly, creating word-of-mouth communications at the speed of light. This area is expanding and changing so rapidly that it's difficult to manage, yet impossible to ignore.

The Social E-volution

Anyone who has ever taken a basic marketing class has come across the Four Ps of marketing: price, place, product, and promotion. These are the core principles of marketing: *Price* represents the pricing strategy for a product or service. *Place* denotes distribution. *Product* defines the product or service. *Promotion* encompasses sales and advertising. These four components constitute the basis of marketing strategy commonly known as the marketing mix. While these core principles still apply, the Four Ps have evolved into what the authors of this section call the Four Es: enticement, exchange, engagement, and experience.

Enticement is the seductive phase of social media. It's the component that draws the consumer in—the friend, the message, the content, the post, or the tweet. *Exchange* is the process or encounter by which a consumer is led. *Engagement* is the execution of the transaction. *Experience* is the overall impression or evaluation of the entire process. The Four Es occur instantaneously and just as with any consumer process, they face the risk of losing

87

a customer anywhere along the line. Today's customers are empowered and encouraged to report on every experience they encounter and are much more likely to do so than ever before. If an experience isn't great, customers may never come back and, possibly more troubling, they may tell all their friends not to come back. A customer scorned may be a social network abandoned.

The evolution of marketing in a social media environment requires end-to-end strategic thinking. Organizations must deliver on their promises or be held instantly accountable. The opportunities for using social media to market transportation are evolving at an exponential rate. Not long ago, computers and e-mail accounts were rationed out only to top management. We are now in an age of instant information with infinite access. Social media marketing that successfully utilizes the Four Es will help mitigate customer problems, manage customer expectations, meet customer demand, and maintain the customer base.

The First E: Enticement
Enticement is all about obtaining a consumer's interest. With thousands of platforms and millions of messages competing for consumers' short attention spans and limited time, how does an agency gain a foothold in the burgeoning social media market? The first step may simply involve listing the organization's home page on printed materials; from there a customer can find links to the agency's Facebook page or blogs. As transportation agencies become more sophisticated and more social media savvy, the opportunities to capitalize on the advantages of social media grow.

By socializing a traditional marketing or advertising campaign, an organization can drive users to its website. From there, visitors can access relevant information or jump to one of the agency's social media channels to interact with other customers as well as members of the agency's own staff. Allowing customers to post about a great new service or product lends immediate validation to others who will be more apt to try something if members of their social networks recommend it. Amtrak uses its Facebook page to encourage passengers to comment about their rail travel experiences, asking riders to share their New Year's Eve train travel photos, for example, and then click a link for more service information.

The Second E: Exchange
Exchange begins with the plan to bait the hook. It can involve the introduction of a new project or program and a simple invitation to respond. It can be a campaign with a social media element. It can be a blog post

announcing a new project or program with a link to click for responses. It can be a video describing a how-to process and a request for comments.

The Third E: Engagement

Engagement is the plum of the social media tree. There is much debate among social media pundits about the definition of engagement. Lee Odden, a public relations professional and CEO of Top Rank Online Marketing, said, "Linking, bookmarking, blogging, referring, clicking, friending, connecting, subscribing, submitting inquiry forms, and buying are all engagement measures at various points in the customer relationship" (Falls 2010). The Social Media Performance Group (2011), a web analytics firm, defines engagement as, "... the proportion of visitors who participate in a specific marketing initiative by contributing comments, sharing or linking back."

For the purposes of this section, engagement is defined as interaction with an audience. It can be based on quantifiable interfaces with users and specifically measured by responses, click-through rates, hits, and similar metrics for each social media platform used. In other words, engagement is based on the numbers and types of social media messages, the quantities and types of audience responses, the locations or sites where the communication took place, and who said what. (A longer discussion of social media engagement can be found in Chapter 6.)

The Fourth E: Experience

The fourth E is the overall impression that a consumer takes away from an interaction. The experience phase involves continued interchanges with the first three Es. Enticement continues and may be refined to reach an even tighter target market. Exchange persists with more interesting interactions that enhance the identity of the campaign. Engagement continues to expand user numbers and build a reputation on the network. The final experience is the end result of the dynamic relationship among the first three Es. Although it is the final phase, experience continues to work with the three other Es as social media evolves and improvements or additions to content are made.

Success Stories

Some transportation agencies have found innovative ways to integrate social media into marketing campaigns that fully encompass the Four Es and have succeeded in enticing ridership, exchanging with users to brand their services, and inviting customers to engage with and experience their systems. Here are some examples.

Iconic Social Media Marketing

The Los Angeles County Metropolitan Transportation Authority (Metro) has successfully used the four Es of social media in a number of creative ways. Efforts include an award-winning campaign that took tongue-in-cheek jabs at the city's overcrowded freeways via YouTube videos. The short productions featured trips taken by a pretty employee posing as Los Angeles Metro's Miss Traffic. Here we can see the first E, enticement, in operation. With a wink at beauty pageants, Miss Traffic traveled in tiara and sash and escorted the viewer on trips to hip destinations using transit, bicycle, and other non-auto alternatives (Figure 4.1).

Miss Traffic's trips served as Metro's bait on the hook and customers were invited to participate as the second E (exchange) occurred. Videos posted on the system's blog garnered thousands of hits and comments by viewers, and triggered the third E, engagement. Metro used the Miss Traffic icon to entice new ridership and brand the agency as an innovative and trustworthy source for economic growth, culminating in the fourth E, a successful experience.

The Washington State Department of Transportation (WSDOT) used this approach to engage citizens in a major project to dig a tunnel under the City of Seattle to replace the SR-99 Alaskan Way Viaduct, a double-deck highway along the downtown waterfront. WSDOT held a contest to name the massive tunneling machine that was being assembled in Japan. The contest was open to students from kindergarten through high school, and the proposed name had to be female and have significance to the state. The winning name Bertha was a nod to Bertha Knight Landes, who was elected mayor of Seattle in 1926.

WSDOT quickly established a Twitter account for @BerthaDigsSR99, and she sent her first tweet from the factory in Japan before making the long journey to Seattle: "So nice to finally have an identity. Maybe now the passport agency will take my application" (WSDOT 2012). "The next generation of engineers is in our classrooms right now," said state transportation secretary Paula Hammond. "Letting students name the machine and providing an opportunity to follow Bertha on Twitter is a great way to engage them in this historic project, which is an engineering marvel."

Whether it's a massive machine preparing to tunnel under a city or a fictitious beauty queen encouraging the public to "Miss Traffic," icons, personas, and symbols can be used effectively to communicate complex

Figure 4.1 With tongue in cheek, Los Angeles Metro's Miss Traffic escorted viewers to interesting destinations via videos that were posted on the agency's blog. (Courtesy of Metro® 2013 LACMTA.)

messages, attract specific target audiences, or reflect an organization's brand. When the concept is effective, the applications in social media are endless. An icon becomes the spokesperson or the "enticing" voice of an organization and can report on specific offerings or events. These personalities can highlight what is available, describe their experiences, generate interest, and interact with the public. If done well, they create a following of consumers eagerly anticipating the next post and ultimately, an experience that customers will enjoy and want to pass on to their personal networks.

Social Media Promotions

Social media and promotions go hand in hand. Transportation agencies commonly use promotions to attract new customers, market specific services, or communicate special deals. For example, when Southwest Airlines announced plans to utilize Akron-Canton Airport (CAK), airport marketing staff created a "social countdown" to generate buzz. They used Facebook, Twitter, and YouTube to build excitement and establish a strong connection between the Southwest and CAK brands (Cecconi 2012). (See Chapter 3 for an interview with CAK marketing staff.)

Not only can an agency easily use social media platforms to send out notices of specific promotions, but the level of engagement can also be tracked easily and reported via response rates and sales data. The key is developing promotions that are relevant to potential users without creating unrealistic expectations. Indeed, one risk is launching a social media promotion that is too successful. Some organizations have had to shut down promotions after only a few hours because demand exceeded fulfillment capabilities or an unexpected number of orders overwhelmed the system.

In May 2011, the U.S. Centers for Disease Control and Prevention made a lighthearted reference to a "zombie apocalypse" in a blog post designed to encourage people to prepare for hurricane season. The page received 60,000 hits—far more than the average of 10,000. The unexpected traffic crashed the server where the blog was hosted (Reuters 2011). While growing demand is typically a desired outcome, agencies should make sure they can fulfill their promises. Word travels fast in the social media world and dissatisfied customers are quick to share their frustrations when their expectations are not met.

Collaboration, Contests, and Competition

If structured properly, social media may be an exceptional tool for engaging customers through collaborative activities, competitions, and contests. For example, the Jacksonville (Florida) Transportation Authority (JTA) invited its customers to name the agency's new smart card. Riders were asked to submit suggestions on JTA's Facebook page or by e-mail. After receiving enthusiastic suggestions for a new name via Facebook, including "Manatee," "JAX Card," and "Jaguar," the agency ultimately settled on "STAR Card" (Jacksonville Transportation Authority 2011).

Similarly, the Washington Metropolitan Area Transit Authority used Facebook to ask riders to help name the new connection between the Farragut North and Farragut West Metrorail stations (Hendel 2011). The Arizona Department of Transportation keeps constituents engaged through its "Where in AZ?" game. The agency posts pictures of obscure Arizona locations on its Facebook page and invites readers to guess where the photos were taken.

On-line video contests represent another way for patrons to publicly showcase their creativity while also enabling a transportation organization to create low-cost promotional materials. Customers receive acclaim for publishing award-winning videos and the agency benefits again from the trust engendered when customers engage with one another about good service or unique programs they experienced. Agencies such as Los Angeles Metro have used video contests to their benefit. Metro's Transit Flicks contest invited riders to submit short videos showing what they liked best about riding buses and trains in Los Angeles County. Metro was so pleased with the winning video entry that the agency later hired the same crew to produce more work. Contests and competitions such as these showcase the enticement, exchange, and engagement and lead to a positive experience (the fourth E) for all.

LIKE! and Other Blatant Endorsements

Another way to encourage engagement with social media is to seek out relevant endorsements. Celebrities or well-known local figures who tweet or post about their good experiences can often influence potential customers to try a service or product. Celebrities who tweet lend credence to a message. When a substantial number of celebrities tweet, the message often goes viral because of their large followings. Endorsements are common forms of promotion, and social media has the potential to forge a strong and effective link between endorser and endorsee.

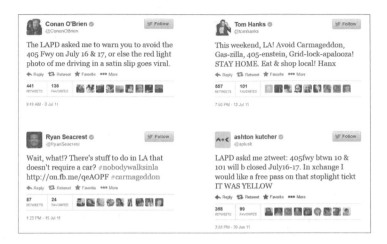

Figure 4.2 In July 2012, local celebrities tweeted about upcoming ramp closures on the Interstate 405 freeway. (Screen captures from Twitter.)

For example, during the 2011 and 2012 closures of the Interstate 405 freeway in Los Angeles for a construction project, local officials used numerous strategies to warn area residents to stay off the roads. Although public support was lukewarm at first, someone dubbed the closure "Carmageddon" and the catchy moniker started an exponential wave of social media. (Metro's full Carmageddon social media campaign is described later in this chapter.) The closure spawned viral videos, Facebook posts, and Twitter updates that captured the world's imagination as followers of the events waited to see whether a massive transportation debacle would unfold. The Los Angeles Police Department solicited local celebrities to tweet about the closure and received responses from Ryan Seacrest, Marlee Matlin, Tom Hanks, Ashton Kutcher, and Conan O'Brien, among others. Figure 4.2 shows some of their posts.

Hiring Big Guns

Transportation organizations are continually finding new and more sophisticated ways to use social media applications to communicate with their followers. Agencies are geo-targeting consumers with pinpoint accuracy with Facebook advertising for high-level projects. The Los Angeles Metro used this strategy as part of its efforts to publicize two separate highway closure campaigns: the I-405 Wilshire ramp closures and the aforementioned Carmageddon closure. Metro purchased premium Facebook ads that were very specifically geo-targeted to customers in the areas of the closures. A premium ad is the sole message delivered on a Facebook user's page, and although the ads come at high prices, they are much more effective than the lists of ads that most Facebook pages feature.

For example, the Metro ad for the Wilshire ramp closures received a click-through rate of over 0.32%, which is estimated to be more than six times the 2012 average for Facebook. Ads for both the Wilshire ramps and Carmageddon led users to relevant Facebook pages that offered more information about the closures along with opportunities to comment or repost the newest project-related information. With this interactive exchange and engagement, it was not long before the social media went viral.

Avoiding Potential Pitfalls and Social Snafus

For all its advantages, using social media for marketing can create challenges for transportation organizations. Following are some potential pitfalls and examples of how transportation agencies got their projects right.

Flash Mobs and Transportation

One aspect of social media that may be of particular concern to transportation organizations is the potential for instantaneous formation of flash mobs. Social networks offer the power to encourage large groups to amass, and they can overwhelm a transportation system easily. In Los Angeles, a midnight bicycle club would tweet out its itinerary to thousands of followers. The festivities typically included visits to several nightclubs throughout the evening but always culminated with a ride home on a local rail line.

In Santa Monica, a popular musical group used social media to attract thousands of fans to one of its performances. When the concert turned unruly, crowds of people rushed to buses to make their way out of town. Even the best-prepared transportation organizations will find it difficult to adjust to hundreds of bicycle enthusiasts simultaneously arriving at a rail station or thousands of patrons attempting to board 40-passenger buses. While not every eventuality can be predicted, transportation organizations need to be aware of the potential impacts of social media campaigns, prepare for the crowds they may generate, and work to resolve potential issues before they become major problems.

Attaining Critical Mass is the Key to Success

In today's marketing environment, it is not uncommon for agencies to incorporate Facebook pages, Twitter accounts, YouTube channels, Pinterest boards, and more into their marketing mix. These types of media have begun to replace print materials such as newsletters, flyers, posters, and brochures (although traditional stakeholders often expect those as well). From an agency view, these social media tools are appealing because they are often perceived as being free. But savvy marketers realize social media requires resources to both provide content and also to monitor online response.

Abraham Lincoln said a house divided cannot stand. Whether or not Honest Abe could have envisioned social media, his remark is still relevant today. An organization that maintains multiple social media outlets may have great difficulty making them all relevant or achieving a critical mass of viewers. When agencies first began experimenting with social media, they expended extensive resources to produce a great variety of social media channels that went largely unseen. Videos were produced, Facebook pages were established, and Twitter accounts were created in the name of marketing and communicating every individual project. One illuminating aspect of social media is that everything can be tracked, traced, and trended. The tale of an amazing social media campaign that "everybody loves and is responding to" cannot hold up to scrutiny if the metrics reveal only 74 friends, 19 followers, and 38 views on YouTube.

While it may be important to reach key individuals on local issues, from a marketing perspective it is more important to get the word out to a wide audience. The social media campaign associated with the Carmageddon highway project discussed later in this chapter is just one example of transportation communications achieving critical mass via social media.

Without a major event like a freeway closure to generate views, achieving critical mass means consolidating and focusing efforts. Individual pages and accounts rarely make a significant impact. As tempting as it is to target market segments within market segments, it is far more beneficial and effective to pool resources and approach social media from a strategic and collective perspective—to brand efforts and build mass by combining messages and cross-populating issues. With those strategies in place, agencies have a far better chance of creating a social media campaign that will capture public interest.

Social Branding: Infinite Opportunities, One Brand
One of the mixed blessings of social media is that almost anyone in an organization can participate. On the negative side, any individual can start a social media site and claim to represent the interests of an organization. On the positive side, as previously stated, an organization can pool multiple internal resources to generate highly engaging content that can brand an organization. In fact, social media can exponentially accelerate the entire consumer process. A consumer can become aware, develop a perception, draw a conclusion, consider using, investigate options, complete a transaction, and become a customer—or not—all in a matter of seconds. An organization can gain or lose a customer anywhere in that process. This is why it is extremely important to have all social media experiences within the brand of an organization.

Social media marketing must reflect the intent of the organization and capitalize on its potential. Individuals must be directed, policies must be put into place, and the brand of an organization should come shining through in all social media efforts. When still in the era of print communications, Dallas Area Rapid Transit (DART) maintained a very strong visual brand in terms of fleet, facility, and communication graphics. As the organization ventured into social media, designers carried the DART brand into everything they did. They worked with designers to create illustrations, icons, and images that were consistent with the DART brand.

The agency now presents information in an informal voice that is customer friendly (no jargon or institutional rhetoric) and consistent throughout its social media materials. Printed materials placed aboard buses and trains often contain references to Facebook and Twitter links. A reader will continue to see the same DART branding throughout the entire print-to-digital journey. All DART customer materials fit style guidelines that perpetuate and enhance the brand.

Home Pages versus Landing Pages
If the social media message is informational only and self-contained (e.g., an alert that says a bus is running 10 minutes late) the task is simple and finite. If the goal is something grander, the message requires additional steps. As was emphasized previously, social media marketing efforts follow a multiple-step process involving the Four Es: (1) entice people; (2) exchange with them and get them to seek out additional information; (3) engage with them and drive them to a certain behavior or transaction; and (4) leave them with a good overall experience.

One of the most challenging aspects in social media marketing is directing consumers and controlling where they land. Conventional wisdom is to send everyone to the organization's home page—typically the showpiece of any website and the funnel to most agency information. But home pages can't hold everything.

Larger agencies experience frequent requests to place something "prominently" on the home page, and complying with every request ultimately clutters the home page and confuses consumers (thank the tech gurus for rotating banners and pull-down menus). In social media marketing if one is attempting to close a deal, sending a prospect to a home page may be a kiss of death. The customer has been led successfully through the first three of the Four Es. He or she has been enticed to read or view a social media component, elected to exchange, is ready to engage, and seeks something specific. Instead, he or she is led to something overwhelming and the experience is ultimately negative.

The alternative is directing the customer to an appropriate landing page. The challenge at this stage is for the landing page to do the job: it must be compelling and satisfy the customer's interest. This is by no means an easy task. Often times an engaging social media post ends up leading a customer to a non-engaging, non-responsive landing page or back to a cluttered and confusing home page. When using social media for marketing, the entire end-to-end process needs to be planned carefully, well ahead of customer engagement, to fulfill the promise. If efforts fail, the other (negative) side of that double-edged social media sword will come slicing down with a barrage of e-mails and scathing blog posts that other customers can read. The prospect to market the product or service successfully will be lost.

Sourcing Information and Providing Content
Sourcing information is another challenge. Social media constantly requires new information and often the most important information comes from deep inside an operating group or accident response team. The information may be very technical, premature, or even present liability implications. For this reason, it is very important to filter or edit what messages are pushed out for public consumption. One solution is an information and style template that provides guidelines and standards.

This solution can be achieved in various ways, e.g., by assigning a single content editor; creating scripts for front line personnel to use; or assigning and training individuals within critical areas. Another reality of instant information is that it can (and will) be rebroadcast very easily with little or no control. For that reason, getting it right the first time is critical. The bottom line is that the information has to be correct, concise, and clear. Transportation organizations that design end-to-end information systems are in much better shape to convey appropriate and accurate information in real time.

Surface Speak and Surface Smarts

One of the byproducts of social media is the superficial exposure of people to vast amounts of information. Consumers continually get information but may never dig below the surface or dedicate themselves to understanding an issue fully. Marketers and communicators have to rethink their messaging to cope with and compensate for what can be termed "surface speak" and "surface smarts." Surface speak is the stream of headlines that everyone is exposed to on a regular basis and surface smarts is the level of information that people actually take in without really reading the full story. A consumer may be aware of a specific issue but lack any knowledge of the details.

This becomes a concern when communicating those extra details is key to the success of an endeavor. For example, when express lanes opened on highways in Atlanta and Los Angeles, people had a surface knowledge of the events but lacked the detailed information and equipment required to actually use the lanes. The results were less than optimal and in the case of Atlanta, were actually hostile.

Marketers have to rethink how they communicate and make sure that critical information is direct and close to the surface because social media users have been trained not to delve into a subject too deeply. Even when a customer does click through to more information, the chances that he or she will read through to the third paragraph are extremely remote. Again, this illustrates the importance of content management and the ability to make a succinct point either via a social media headline or early in the click-through process. Social media is set up ideally for surface speak, and if the content is coherent and concise, the message will be relevant, memorable, and useful.

"Liking" Social Media

Transportation organizations create Facebook pages because they can market their services and projects on the number one social media site. Establishing a Facebook presence extends an agency's mission. An agency may be neglecting communication opportunities if it does not engage with customers on Facebook. All social media including Facebook provide services that give an agency immediate feedback for system improvements and changes, and can also serve as barometers of a system's entire customer experience.

Listening carefully to user responses and suggestions is a key mechanism for keeping in touch with the needs and desires of an agency's customer base. The great advantage of social media is that it is very easy to check its pulse. If potentially damaging information surfaces, it is increasingly easy to see whether it gains traction with the public. For example if an unfavorable news story doesn't gain any comments on social media, chances are the issue will dissipate quickly. In a political environment, transportation professionals have to be sensitive and consider the impacts on elected officials. However, from a marketing standpoint, if little or no "chatter" appears on social media channels, the worries about reputation damage abate.

One more important concept is not to make a bad situation worse. For a quick pulse on any issue, it is easy to check social media channels, but the magnitude of the Internet must be kept in mind. For example, an issue garnering ten negative comments may appear to be a concern, but ten is a relatively unimportant number of responses in comparison to more than two thousand concurrent postings about a local sports figure. The public has its priorities! In cases such as these, a good content manager can help determine what the comments mean and how best to respond. (Chapter 6 has a detailed discussion of how transportation agencies can address online criticism.)

Trending as Success Metric
Marketing metrics have always been difficult to quantify. Because customer creation is typically a multi-step process. a direct correlation between the allocation of marketing resources and the return on investment is difficult to determine. Introducing social media into the marketing mix has complicated such issues even further. While views and visits can be identified, their impacts cannot.

While yesterday's marketers were looking for reach and frequency, today's marketers look at what's trending. Trending has become the new metric for success. If your company, product, or service is creating a buzz on social networks, the chances are that whatever ignited that buzz worked. Carmageddon is just one example of a marketing strategy that caught fire in social media circles. The thought of closing down one of America's busiest freeways for a weekend captured clicks and views around the globe.

Marketers want their products and services to trend—and trend positively. Commercials, advertisements, and promotions, if handled properly, can create immediate momentum as they launch into cyber space. A campaign that goes viral garners a far greater return on investment than one that is limited strictly to the media buy. Today, many commercials are designed to live long after their airtime or they are created to entice viewers to look further into what they offer.

It is very common for advertisers to ask specifically for viewers to like, visit, or subscribe to their pages or sites. The difficulty is in valuing the social media buzz. Just as yesterday's marketers would tout impressions, awareness, or recall of a campaign, today's marketers look to value social engagement to convey the impact of a particular campaign. The ability to use social media trending as a metric is becoming more mainstream. (See a longer discussion of social media metrics in Chapter 6.)

Avoiding Social Blunders (and Blinders)

Transportation agencies that don't want to miss the boat (or bus or train or plane) are scrambling to join in the social media fray. However, firms that rush to open social media accounts often make social media blunders that may be avoidable. Marketing strategist and prominent blogger Michelle Golden (2011) advises that content sharing must take place along with relationship development and that the two aspects go hand in hand. She warns of misunderstandings about the definition of social media—the failure to understand that social media is built on relationships.

Often transportation agencies mistakenly employ social media tools for one-sided content delivery. By allowing back-and-forth conversations and talking with—instead of talking at—the public, social media marketers can gather large audiences that trust their message on a variety of sites, thus avoiding social media blunders.

In addition to blunders, agencies would do well to carefully avoid social media blinders. Social media evolution has been extremely fast-paced. Noor Al-Deen and Hendricks (2012) explore its evolving impact as they review use and effects of social media for various purposes including education, communications support, politics, legal arenas, and ethical platforms. With the rapid proliferation and adoption of social media, the authors insist that social media must be addressed in nearly every aspect of business.

Some consider the transportation industry an old-school operation that still gets much of its news and information from printed materials. In fact, the percentage of workers in the transportation and warehousing

101

sector aged 55 to 65 increased by about 33% between 2000 and 2007, and the proportion of workers over age 65 increased by about 15% (Sweet and Pitt-Catsouphes 2010). While age is not a definitive indicator of social media use, young adults are more likely to adopt social media than their older counterparts, irrespective of industry. The aging transportation workforce suggests that some industry personnel may be unfamiliar with using any of the developing web technologies or reluctant to try them.

Social media is changing the ways in which all businesses communicate, including those in the transportation industry. Rebel Brown (2011), an Internet marketing expert and blogger says, "It's time to shift our marketing approaches from traditional me, me, me-focused campaigns and promotions to interactive conversations and relationships. That's where we'll find our marketing power in today's world." Social media networking is now permanently woven into the fabric of our nation's consciousness and has been embraced in a multitude of ways. The transportation industry's need for getting up to speed with current practices in social media marketing is a must if the industry wants to survive, grow, and meet customer expectations in today's tech-savvy world. Brown (2011) puts it this way:

> Everything shifted and is continuing to shift—faster and faster. If you're using the same basic business and marketing approaches with a few updates here and there, you're wasting time and dollars. Successful executives consciously and continuously adapt their products, sales and marketing think, incentives, and more to intersect their markets as they evolve.

With each new technological shift the youngest members of the workforce often possess the greatest skill, ability, and attitudes to utilize the evolving applications. Social media is no exception. In a time when older members of the workforce may still be using cork bulletin boards, savvy students are utilizing *online* bulletin boards and other social media platforms with ease—and much better results. Social media use is second nature to younger members of the workforce. The key is to merge the technological know-how of the young with the maturity and experience of older workers. This merging of new technologies with existing wisdom in the area of content delivery can achieve maximum results.

Marketing to Tomorrow's Customers

Tomorrow's customers offer a new set of marketing challenges. The social aspect of social media will evolve and become even more targeted and tailored. Tomorrow's customer will access information directly from monitoring systems that automatically identify, customize, and transmit specific and useful information to tablets and mobile devices. Machine-to-machine (M2M) communications will link a person's whereabouts with traffic systems, event calendars, retail establishments, and predetermined social networks to offer customized information specifically tailored to his or her demands. QR codes, printed materials, and company websites will be replaced with custom micro sites tailored specifically to each user. Connectivity, information, and exchange will happen instantaneously. Individually targeted content will become the commercials and social networks will be automatically identified and defined. Tomorrow's customers will receive information and trip data directly from the systems that monitor information. They may enjoy a live chat with a customer service system that knows the consumer's exact location and also knows all associated traffic conditions transmitted directly from traffic sensors or roadside cameras.

Every phone will be smart and automatically provide relevant personal information to social networks in exchange for relevant transportation information that may be fed directly into a vehicle or ticketing program. Retail systems will automatically prompt for tweets, then process and socialize as needed. Digital signs defined by shorthand icons, prompts, or vocal instructions will appear on consumers' mobile devices.

Yesterday's transportation customers relied on media networks for information and enticement. Today's customers exchange information and engage transportation through social networks and rely on the availability of instant information. Tomorrow's transportation customer will have the world automatically delivered to the palm of his or her hand, exactly tailored for the ultimate social experience. Tomorrow's marketers need to prepare today.

A CANTANKEROUS NOISE: SOCIAL MEDIA AS AN ADVOCACY TOOL
Andrew Austin and Ashley Robbins

Before starting this segment, let's clear the air on one item of business. Social media is an important tool in the world of transportation advocacy, but if you rely on it alone, you will lose.

What!?! You may say, "I thought this was a chapter about advocating for progressive transportation through social media." Fear not; when social media is used well, it can and will assist in winning. When online organizing is designed to complement and enhance your coalition building, lobbying, media activities, and old-school grassroots organizing campaigns, it works. But if you replace these core advocacy strategies with social media alone, you will wake up one day with lots of followers, thousands of likes, but no real power for your cause or results in the community.

This section draws lessons from our on-the-ground (or on-the-web) experiences. We are advocates who hail from Seattle and Atlanta, respectively. We are two people among thousands working in the transportation advocacy and organizing space across the country, and there are many others to learn from.* This segment shares best practices and lessons learned and offers a glint of inspiration on how to use social media to build a better day for walkers, cyclists, and transit riders in every community. While our experience focuses on transit, bicycling, and pedestrian programs, the lessons can be applied to all modes of transportation at any level.

Working toward Livable Communities

Despite historic shifts in public demand and desires, transportation choices are still limited in many parts of the United States. Tireless advocates, elected officials, and agency employees are working to move away from 80 years of the car as king. Former U.S. Department of Transportation Secretary Ray LaHood said it best when asked to define a livable community. "It's a community where if people don't want an automobile, they don't have to have one" (Schor 2009). Despite increased investment in walking, cycling, and transit infrastructure, many Americans still cannot make that choice even if they want to, and they remain isolated from work, school, and their families because they cannot or choose not to drive. The situation is getting better, but as a country we are nowhere near there yet.

Since the 2008 recession, transit agencies and their riders have faced historic service cuts, fare increases, and route abandonments nationwide. Simultaneously, the public demand for transit is at record levels. Americans are demanding more transit, not less. According to the Center for Transportation Excellence (www.cfte.org), between 2002 and 2012,

---

* For a list of transit advocacy organizations, see http://americansfortransit.org/transit-organizing-directory/ and for active transportation organizations, see http://www.peoplepoweredmovement.org/site/index.php/site/memberservices/C530.

over 70% of local ballot tax measures to save service or expand systems succeeded in rural, suburban, and urban communities. Ridership grew more than 3% nationwide in the first 6 months of 2012 (APTA 2012). In September 2012, a national poll indicated 64% of Americans believe their community would benefit from expanded and improved public transit systems and 68% support more local investments to pay for it (Perk 2012).

On the bike and pedestrian side, despite momentous gains, many U.S. cities remain uncomfortable, hostile, or outright dangerous for those traveling on foot or by two wheels. Better cycling and walking infrastructure is on the rise, but is far from ideal, especially in secondary cities, the suburbs, and more rural areas. With the growing obesity epidemic, global environmental realities, and life or death safety challenges, many are working to ensure that attractive and safe active transportation infrastructure in conjunction with great transit services will be available for everyone.

The dichotomy between the transit funding crisis (and the overall lack of funding for active transportation) and the widespread demand for more transportation choices makes the job of progressive transportation advocates even more timely and important. It is not an overstatement to say the future of local communities, the country, and global environmental health is counting on individuals in localities across the U.S.

Where Social Media Works

As already mentioned, social media alone will never win the transit–bike–walk debate. However, social media is an easy-to-use and cheap tool that will help achieve victories and it is becoming an increasingly important component of grassroots campaigns. In the metaphor of the proverbial advocate's toolbox, social media and online organizing constitute the tool belt that allows the use of all of the other advocacy tools with greater potency, ease, and efficiency. Social media is the common communications connector of advocacy campaigns—the neurological system pulsing between all the other campaign components.

First and foremost, social media is powerful when used to engage, educate, and organize a base of support. If the ultimate goal of an advocacy campaign is to win funding or a policy debate, it usually involves convincing an agency, or more likely politicians hired by their constituents, of the value of a request or "ask." The ask may be an idea, money for an idea, or both. If a room full of individuals demanding funding to stop bus service cuts consists of three avid "transit nerds" or four diehard cyclists asking for a bike boulevard, the group will not move the public debate. Best practices show the need to pack hearing rooms, overwhelm media, and create a cantankerous noise to win in the toughest political transportation fights. When done well, online organizing can drive in-the-flesh human beings to attend events, volunteer actions, and public hearings. It can amplify the noises and voices of the individuals organizing at those gatherings and keep those who couldn't attend in the loop. Social media serves as a crucial factor for building real people power for a better transportation future.

It should be said: social media will never reach everyone. While the bus riding 90-year-old grandmother may have finally signed up for Facebook so she can see pictures of her granddaughter, Facebook is unlikely to be an avenue for her to read the news or hear about events. To convince her to attend a hearing to tell her city council member to fix her broken sidewalk or not eliminate the bus line, a knock on her door or a handwritten letter will likely be needed.

Online organizing works best with a critical mass of plugged-in people—young adults, urban America, school campuses, compact commercial districts where most employees don't drive to work, and communities built around high capacity transit lines.

What do most urbanites do while sitting on a train? How do commuters spend their time on their suburban express buses? Nearly 70% of commuters use their cell phones while in transit. They are texting, reading e-mails, tweeting, and generally killing time while riding (Frei and Mahmassani 2011). Ninety-two percent of those under the age of 30 report using social media; so do 73% of people under 50 and 53% under 65 (Brenner 2012). Reaching transit riders through social channels will help organizers recruit and motivate them to move beyond being passive riders to become engaged advocates.

Educated and tech-savvy urbanites are getting their news from Twitter, planning their evenings through Facebook, and sending work e-mails between those activities. And they can handle all these tasks from seats on a favorite transit route that gets them to work or school every day. Social media tools facilitate the educations of many people who live car-free (or car-light), create dialogues between advocates and their supporters, and help build long-term advocates for transportation choices. Here are two stories of doing just that.

Advocating for More Transit Deep in the Heart of Georgia

In July 2012, Metro Atlanta residents were asked to vote on a referendum to fund transportation. The referendum, originally authorized under the

Transportation Investment Act of 2010 (TIA), was part of a statewide campaign known as T-SPLOST (Transportation Special Purpose Local Option Sales Tax). Citizens in each of Georgia's 12 economic development regions voted on a 1-cent sales tax to invest in a locally defined list of transportation improvements. While the T-SPLOST referendum was ultimately unsuccessful across the state, the campaigns leading up to the vote were able to capitalize on social media to sway the project list in favor of transit and pass the referendum in the urban core of the region.

Setting the Stage
A grassroots, transit advocacy group called the Livable Communities Coalition (LCC) engaged in a two-part campaign leading up to the ballot. In the fall of 2011, the first phase, the Fair Share for Transit campaign, sought to influence the creation of the project list that would go before voters the following July. Over the course of 3 months, a roundtable of 21 elected officials was responsible for culling the list from 417 projects totaling more than $29 billion to a list with a $6.14 billion price tag. The goal of the Fair Share for Transit campaign was to secure 40 to 60% of the funds for transit projects. An active campaign ensured that transit was awarded 52% of the available funds.

The campaign's influence over the projects included on the list hinged on gaining the support of the advocacy community. To garner support, the campaign placed a heavy emphasis on social media, including an active Facebook page, Twitter feed, e-newsletters, and blog. This robust social media strategy was able to turn followers into active voices at public meetings to speak in favor of transit projects. Hundreds of volunteers attended public meetings wearing green shirts and passed out "Fair Share" stickers, "More Transit" buttons, and information on the transit projects. More volunteers came out as the campaign moved into the "Fast Track Forward" phase to build support for the referendum by making thousands of penny lapel pins, calling voters, and passing out maps of the transit projects to commuters.

Building an Online Community
The key to the successes of the Livable Communities Coalition's social media initiatives was to capitalize on previously established networks including the Twitter feed of one of the authors of this section, a well known transit advocate in the Atlanta community. Using the handle @CCTGirl and the hashtag #TIA2012, she sent out numerous tweets in support of transit improvements:

@CCTGirl: In case you forgot, Atlanta ranks 9th in the nation in congestion and 48th in the nation in transportation spending. #TIA2012.

@CCTGirl: The city of Atlanta will pay $915m into the #TIA2012 and will get $1.5b back when MARTA state of good repair monies are included.

@CCTGirl: The city will receive $94m in local funds to use at their discretion. #TIA2012. Of the 86 e-mail comments received 60 pertained to bikes.

Building a community and a following online takes time, and it is important to establish credibility and trust among social network channels in order to turn an Internet follower into a real-world advocate for hearings or volunteer events. Creating Twitter and Facebook accounts 6 or more months in advance will enable organizations to begin to gain followers well before their efforts are needed.

Two of the quickest ways to establish credibility and make social media networks the go-to sources for information is to send out relevant news pieces and to live tweet meetings, presentations, and public hearings. Not only do these measures draw a following, but they also serve to educate the community on an organization's issues and perspective.

As news agencies retweeted campaign tweets from speaking engagements and public hearings, the campaign's social media channels grew. The keys to capitalizing on these events are to (1) push the information out quickly, focusing on key points and quotes from officials and the community, (2) use hashtags to make the tweets searchable, and (3) use Storify to summarize event tweets. If a hashtag has not been established for a referendum or campaign, organizations should quickly create one and use it for the duration of the campaign. Consistency is key.

The most crucial component of any successful advocacy social media campaign is to have a personality—find the group's voice and have fun. A successful social media campaign builds a strong community of followers who regularly engage and then act. At its core, social media is about connection, and in a world full of thousands of Facebook pages, Twitter feeds, and an overstimulated online audience, being identifiable and relatable is the number one way to get people involved. Always keep in mind the target audience for each social media channel; knowing the audience is crucial to making a personal connection with them.

Forget formalities and use pictures of people. The TIA was hotly contested within the community and the LCC came under attack on Facebook and Twitter until organizers switched from using the campaign logo to a photo of the LCC social media coordinator. Showing the coordinator's face helped humanize her interactions online. Negative and accusatory comments became cordial debates and the woman filling the position became easily recognizable at meetings.

Creativity is another way to grow excitement around a social media campaign. As Facebook friends post likes and tweets are retweeted, awareness around the campaign is heightened. The LCC campaign engaged and educated followers using infographics, historical photos of Atlanta transit, an inspirational video composed of movie clips, and Internet memes, which are popular phrases or images that incorporate a cultural or social idea and typically spread quickly across the web. Having constant content that is both informative and entertaining translates into momentum that can then be used to make requests of an online community.

The goal of any advocacy campaign is to turn likes into action. Once a social media campaign is rolling with an established, credible, timely, entertaining, and recognized voice, organizers can push out meeting notices and event invitations. Facebook pages can display photographs of volunteers working; tagging them in posts can stroke their egos and let their friends know what the campaign is doing. Making volunteers temporary administrators of a campaign Facebook page enables them to invite their friends to like the organization's page and events while letting them feel ownership of the campaign. Setting up texting trees with a group of dedicated volunteers to ask them to engage with the organization's posts will bump the content up in the newsfeed. And always reward volunteers with pizza.

Lessons Learned from Fast Track Forward

- Like any marketing campaign, social media is most effective when it is identifiable. Building a personal brand allows a campaign to stand alone in the mainstream while energizing the group's base.
- A key use of social media is educating supporters and skeptics. After followers are engaged in discussion and aware of key issues, organizers can ask them to fill chairs at meetings, but they must first create that initial buy-in from followers.
- A social media campaign should be one of the first tasks on an organization's to-do list. Start it early, engage with the base often, and remember that consistency is key.

- Make social media interactions personal; connect with followers, use photographs of staff and volunteers, and do not forget that people are online to find community and human interaction.
- Know the audience. Followers are most likely young and urban minded, and they are not going to understand policy wonkiness or jargon.
- Have fun. Use surveys, memes, images, videos, everything at the organization's disposal to engage and energize the base while creating the impression that this campaign is the greatest thing to be involved in.

Social Media Helps Save King County Metro

Serving the Seattle metropolitan area, the King County Metro system (Metro) is the largest transit agency in Washington State and the eighth largest bus system in the country. Due to the 2008 economic collapse, transit riders in Washington State and the entire nation experienced severe route eliminations and bus cuts were hitting agencies and transit riders. Without stop-gap funding from the King County Council, Metro would have been forced to cut 20% of its service. Given Metro's size, these cuts were the equivalent of completely eliminating the second largest bus agency in the state.

There was an alternative to such draconian cuts, however, in the form of a $20 per year vehicle registration fee, and public hearings on the proposal were scheduled. A targeted, well-planned, and active social media strategy was key to achieving victory. Before the hearings on the car registration fee or "tab," the media was in a frenzy about taxes. News stories displayed headlines along the lines of, "King County Council wants to raise your car tabs," leading observers to think that taxpayers were in an uproar over paying $20 more per year.

That all changed on July 12, 2011. In Downtown Seattle, a wall of humanity stretched six blocks down the city's main avenue waiting to get into the Martin Luther King County Council Building; upstairs the King County Council was holding its first of many public hearings on the future of King County Metro.

The Council chambers, the overflow rooms, and the hallways were packed with hundreds of transit supporters. The news reports estimated 1,000 or more people showed up for the hearing and over 300 people waited in line for hours to testify in support of saving the buses. One pro-transit county staff veteran commented, "In the 30 years I've worked in

Figure 4.3 Hundreds of Seattle Metro riders attended a King County Council hearing to support transit. (Courtesy of Andrew Austin. With permission.)

government I've never seen public involvement like this." Every local TV station covered the event, panning their cameras across the endless lines of people chanting in support of buses with their handmade "Save Our Metro" signs in hand. Figure 4.3 shows some of the supporters.

Due to the fast-paced and nearly manic atmosphere, Twitter was the primary social medium used around this event. Supporters, whether present or not, were able to engage in a robust discussion about how important their bus service was to them and people they knew. The primary hashtag of the night was #saveKCMetro and it went viral to the point that it was a trending topic on Twitter across the region. Some sample tweets using the #saveKCMetro hashtag are shown here:

@t4wa: If you support the @KCCouncil saving @KCMetroBus today, right now please tweet: #SaveKCMetro

@seattlest: Many more testimonies from outside of Seattle at this hearing. Buses = not just a "city" issue, but a COUNTY issue. #SaveKCMetro

@mshannabrooks: 95% of King County's bus riders have a car available for their use. They choose transit, and support a fee. #SaveKCMetro

Everyone told a different story after that night, from people tweeting at home, to bloggers, to major TV channels. Reporters could not ignore so many people demanding that the county council step up to save services so crucial that most citizens couldn't imagine would ever be in jeopardy: their buses. The narrative in the weeks ahead did not focus on the proposed temporary $20 vehicle license fee, union pay, or supposed government waste. Instead, media focused on the story of thousands of individuals who rely on and love transit.

Social media alone did not create the outrage and historic public engagement on the local level, but it helped win the campaign. First and foremost social media was used as a tool to engage supporters over the long term, convince them to show up, and make their voices heard. Through Facebook events, Twitter notices, and old-school phone banks, the advocates ensured that hearing after public hearing was dominated by pro-transit voices.

Furthermore, social networking was a crucial tool to keep supporters motivated and frame the conversation within the spheres of larger public discourse. When #saveKCMetro was trending on Twitter, riders going to their evening shift at work or moms at home with the kids could chime in, follow along, and be part of a movement. Twitter allowed all the major decision points and hearings to be live tweeted to the universe along with a narrative showing how important transit service was to the community. The impressive result was that not one single advocacy voice (or handle) shaped this conversation. It was an organic virtual outcry that mirrored the packed physical hallways of the county buildings for weeks of events. Clearly, social media outreach provided an important tool for advocates in helping shape the message around the need for saving transit service. Not only was it used as an avenue to present heart wrenching stories, it was also a way to share talking points and facts with supporters at public hearings.

Finally, social media gave voters, riders, and supporters an important way to engage the transit agency and their elected officials. The transit agency could not adopt the hashtag #saveKCMetro because it was associated with an advocacy movement. Instead Metro created its own hashtag #metrofuture. Through this medium, the agency was able to reach out to riders and thousands of followers and inform them of continuing discussions about the future of their service. The agency's savvy social media team and the staffs of the transit champions in the county played into the virtual discussion around the future of transit in a way that complemented and strengthened the advocates' voices instead of drowning them out.

On August 12, a month after the first public hearing, the King County Executive called a press conference announcing that the council secured enough votes to pass the temporary fee and save Metro. What was first thought to be impossible was accomplished. It could not have been done without organizing the people power that was fueled and fostered by social media and online organizing.

Lessons Learned from Saving King County Metro

- Social media is a crucial tool for capturing existing supporters, honing the collective advocacy message, and driving turnout to public hearings and events. It is an inexpensive (or free) means to reach thousands of supporters quickly, build momentum, and organize people around a common goal.
- Live tweeting public hearings was useful for keeping the public engaged, framing a story for the traditional media, and capturing the electric in-person energy. The story went viral and greatly amplified everyone's voice and democratic participation.
- Coordinated and complementary social media strategies of agencies and advocates can go a long way. The messaging, engagement, and plan should be designed in a way to amplify each party's efforts toward common goals.

A Better Day

Across the country, a strong and growing community of progressive transportation advocates is developing. The common goals and beliefs of this community are easy to identify: they want governments that make transportation policy decisions that prioritize people. As advocates, our most important role is moving the people-oriented agenda forward and social media is instrumental in accomplishing this. While the relevance of social media continues to grow, it is still evolving and its strengths and limitations are still being established. Everyone who has used Twitter, Facebook, or any other online tool has cautionary stories to tell. Some of the lessons gleaned from these tales are

- Don't argue. Some naysayers will never be convinced.
- Overposting doesn't mean a message will be seen by more people; it's more likely to turn people off.
- As soon as anyone figures out the Facebook algorithm to increase the chances that a post will be seen by more people, Facebook will change it. Don't stress about it.
- Don't link a Twitter feed and Facebook page to automatically post to each other. These two media serve different, but sometimes overlapping, crowds. If they both say the same thing all of the time, what would be the point in following both?
- Never underestimate the benefit of flyers, in-person meetings, and phone calls. Traditional forms of organizing will never lose their value.

Social media and online organizing, when done well, serve as crucial tools to help advocates win. They are affordable resources that connect supporters, assist in communicating messages, and help build grassroots power. When advocates for a transportation system that prioritizes people over cars win, communities and their members all experience a healthier, more affordable, and sustainable transportation future.

FROM THE SILICON HILLS TO THE SEA: USING SOCIAL MEDIA FOR FEEDBACK AND COMMUNITY ENGAGEMENT
Jody Feerst Litvak and Jennifer Evans-Cowley

Metropolitan areas across the country are giving careful thought to improving their transportation systems by adding everything from new runways to expanded freeways to new rail lines. The planning, design, and construction processes for such major infrastructure projects are complex, generally requiring extensive analysis and multiple levels of review and approval. These capital investments typically necessitate significant public engagement to fulfill state or federal requirements and help shape a project. Robust public input can also build public support and buy-in for a project.

Traditional outreach techniques usually include a combination of public meetings, project advisory committees, press releases, newsletters, advertisements, and electronic communications like e-mail updates and project websites. Over the past few years, public agencies have begun to experiment with using social media to expand their reach when seeking community feedback for projects, programs, and plans.

This section describes major communication campaigns in Los Angeles and Austin, where officials combined traditional and new media strategies to engage the public in discussions about transportation plans and projects. In California, the Los Angeles County Metropolitan Transportation Authority (Metro) incorporated social media elements into the planning and public outreach process for a subway expansion project. In Austin, Texas, city and regional officials piloted, tracked, and evaluated the use of social networking to build relationships and gain valuable public participation in the city's long-range planning process.

Los Angeles Westside Subway Extension

The long envisioned subway serving Los Angeles' Wilshire corridor, initially known as the Westside Subway Extension, has been approved and construction is expected to begin in 2014. The project is now called the Purple Line Extension. The public involvement leading to this point has been extensive and the use of social media substantially enhanced the results.

Wilshire Boulevard, the city's main east–west corridor, begins downtown and runs more than 15 miles west where it ends at the Pacific Ocean. Sometimes referred to as a "linear" or "second" downtown, it serves some of the area's densest job centers and key destinations including the Miracle Mile (home to many museums), Beverly Hills, Century City, and Westwood (home to the University of California, Los Angeles). Analysis in the Environmental Impact Statement/Environmental Impact Report for the subway showed that more than 300,000 people commute into the Westside area for work every weekday (LACMTA 2012). That exceeds the number of people who live in the area and commute to work elsewhere plus the number who live and work within the Westside.

Wilshire Boulevard is also the area's most heavily traveled bus corridor with more than 80,000 boardings on an average weekday. The western stub of Los Angeles' current heavy rail Purple Line subway extends to Wilshire and Western Avenue, just three miles west of downtown. The new project will extend that rail line an additional nine miles and add seven new stations. Figure 4.4 shows the planned alignment and station locations.

Los Angeles has been discussing how to address mobility issues along Wilshire Boulevard since the early twentieth century when

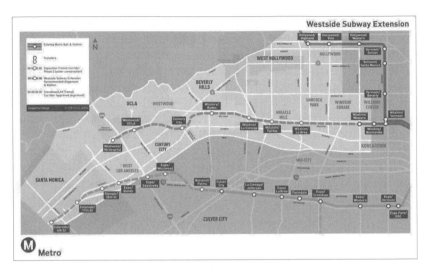

Figure 4.4 Westside Subway Extension will extend the current heavy rail Purple Line from Wilshire and Western nine miles to the west and add seven new stations. (From: Los Angeles County Metropolitan Transportation Authority. 2012. Westside Subway Extension: Final Environmental Impact Statement/ Environmental Impact Report. With permission.)

solutions such as horse-drawn trolleys and streetcars were contemplated and ultimately rejected in favor of motor coaches. Figure 4.5 shows passengers boarding a bus operated by the Los Angeles Metropolitan Transit Agency, one of Metro's predecessors. Subway planning efforts date back to the middle of the last century. Until now, numerous attempts failed to fulfill the vision. The most recent attempt before the current plan started in the 1990s and was stopped for a variety of reasons including technical and engineering challenges, funding problems, and a lack of public support (Taylor et al. 2009).

As the transportation planning and programming agency for Los Angeles County, Metro began the current effort in 2007. Metro began utilizing social media strategies almost from the very beginning. This may seem commonplace less than a decade later, but the Westside Subway Extension was one of the first transportation projects planned under the National Environmental Policy Act (NEPA) to use these strategies. More notably, even today, is how extensively the project utilized social media techniques, how fully incorporated these techniques were throughout

Figure 4.5 Passengers boarding a westbound bus on Wilshire Boulevard in Beverly Hills circa 1960. (From: Los Angeles County Metropolitan Transportation Authority Research Library and Archive. With permission.)

the environmental planning phase, and how they remain so now in preconstruction and continuing into construction.

Latest Planning Efforts

Metro's subway planning efforts followed the requirements of both NEPA and the California Environmental Quality Act (CEQA). While Metro undertook environmental planning, the Federal Transit Administration (FTA) was involved at every step and was the lead agency for NEPA purposes. Environmental planning started with an alternatives analysis (AA) study. In January 2009 the Metro board of directors approved the study and authorized proceeding with a draft environmental impact statement/ environmental impact report (EIS/EIR). In October 2010, the Metro board selected a locally preferred alternative (LPA) for the project and work began on the final EIS/EIR. For CEQA purposes, the Metro board certified the EIR and approved the project in April 2012. The FTA issued a record of decision for the project in August 2012, signifying the end of the federal environmental review.

For much of the middle of the twentieth century, many in greater Los Angeles questioned whether rail transit was necessary or appropriate in a region that grew and prospered through use of the automobile. This debate involved much discussion during earlier subway planning efforts for this corridor. However, the most recent planning effort occurred in a significantly changed environment. From 1991 until this effort began, 17 miles of heavy rail and 51 miles of light rail were opened within Los Angeles County. Commuter rail also began operating in conjunction with the adjacent counties. All these new services certainly created issues and controversies. Nevertheless, by 2007, many old attitudes about the practicality of rail faded as train service moved from a theoretical concept to reality for much of the metropolitan area.

In 2005, Antonio Villaraigosa captured people's imaginations as he made transit expansion and the "Subway to the Sea" one of the focal points of his second and ultimately successful campaign to become the Mayor of the City of Los Angeles. The probability of delivering the Westside Subway Extension and other transit improvements increased significantly in 2008 when a county-wide ballot initiative called Measure R received the required two-thirds approvals from Los Angeles County voters and increased sales taxes for transportation by a half cent. Measure R included an expenditure plan dedicating the revenues for a variety of purposes including the subway and 11 new transit projects.

Community Participation Strategies
Metro's outreach for planning new rail and other projects generally goes well beyond the minimums required by CEQA and NEPA and this practice has usually served the agency well. Planners for the Westside Subway Extension felt an even greater obligation to work with and include the public—stemming in part from the project's history of failed attempts. Planners sought to learn as much as possible from the lessons of their predecessors and allay many of the public fears that foiled past attempts. Planners were further motivated because the extension had higher visibility throughout greater Los Angeles than many, if not all, of the other Measure R projects because it was projected to consume up to 10% of the mandated tax revenues over the next 30 years.

Traditional Strategies
Beginning with the AA in 2007 and continuing throughout environmental planning, Metro utilized all the standard outreach tools of project planners and also employed new techniques. Traditional approaches included:

Meetings — Seventy-one public meetings were held, including requisite scoping meetings and hearings, community update meetings, and meetings of six station area advisory groups. All meetings were well publicized and open to the public. All Metro-hosted public meetings offered simultaneous Spanish translation, Korean and Russian interpretation where needed, and captioning services on request for those with hearing disabilities.

News media — News releases were widely distributed to announce Metro-hosted meetings and publicize key milestones. Metro also provided detailed press kits containing image CDs and other project information to media representatives attending the community meetings. On a few occasions, attempts were made to "pitch" stories to certain reporters.

Mailers, notices, and e-mail — Metro developed and continuously updated an extensive database for the project. Meeting notices were sent by USPS and e-mail and distributed on Metro buses, trains, and at key locations. E-mails were used to release information quickly, or more frequently to serve as meeting reminders, note comment deadlines, and announce project milestones.

Advertising — In advance of community meetings, Metro regularly purchased display advertising in local community papers including those targeting non-English speaking communities in the study area. As the project continued, banner ads in online news outlets and blogs were added to the effort.

Project website — www.metro.net/westside has continued to grow and expand since the inception of planning in 2007. It now houses virtually all the public information that has been produced for the project including myriad reports, copies of presentations, frequently asked questions, fact sheets, and news stories. It also includes a comments page where stakeholders can easily submit opinions or questions and sign up for project updates (Figure 4.6).

New and Social Media Strategies

Changes in the traditional media landscape were well underway by the time subway planning began in 2007. This included the demise and consolidation of various community news outlets. Most significant for this project, however, were staffing changes at the *Los Angeles Times*, which could no longer be counted on to cover subway planning and development. While fewer reporters at traditional news outlets covered local issues, various new media outlets were blossoming.

Figure 4.6 Main website for Westside Subway Extension. Pull-down menus on the right provide links to more information. (Screen capture from http://www. metro.net/projects/westside/. With permission.)

Just to put this in perspective, this was the social media landscape in 2007. YouTube was present but not as pervasive as it is today; likewise, Facebook was hardly the behemoth it is now; Twitter barely existed; most print and broadcast news outlets were still figuring out what their online presence would be; and few if any fully online news sites like *Huffington Post* or *Patch* existed then. This situation presented both opportunities and challenges to subway planners.

Blogs and Online News Outlets
For their first foray into this new world, subway planners concentrated on developing techniques to address the changing news media landscape. While not ignoring traditional media, they focused on identifying, contacting, and developing relationships with bloggers at *LAist*, *Curbed LA*, *Streetsblog LA*, and others who covered planning, development, transit, and transportation issues. At that time and even now, these sites did not have the reach of traditional news media, but they could be counted on

to reach the individuals who were most interested in the project, many of whom were opinion leaders.

Quite often, one of these blogs would be the first to cover some aspect of subway planning. The story would then get picked up by other outlets including traditional print or broadcast media, thereby expanding the reach of the message. Metro also worked with the online news outlets provided by traditional news organizations, particularly the *Los Angeles Times' Road Sage* blog until it was discontinued. Metro made no distinction between online and traditional news sources when responding to queries and continues to monitor the changing media landscape to identify new opportunities.

Beyond working with writers, reporters, editors, and bloggers, Metro actively reviews comments posted online in response to stories. Sometimes a Metro representative participates in an online discussion to answer a question but the agency usually found it more productive to let other members of the public make such clarifications. The information provided in these comments was often seen as more credible since it came from independent sources.

More frequently, when a blog post began to generate online responses, Metro took advantage of the show of interest and encouraged commenters to join the study's "official" public comment process. A Metro representative would join the online discussion to invite those interested in the subway to send their comments to Metro, add their names to the project database, attend meetings, "like" the project on Facebook, or follow it on Twitter. When doing so, Metro representatives always clearly identified themselves as agency employees.

It's important to note that not all news organizations have the same policies about reporting stories and the rules are less clear for online news. The editors at traditional news organizations may, for instance, require their reporters to verify information with two or more independent sources before reporting an event. Online news sites, whether parts of print or broadcast news organizations or solely online outlets, often do not have the resources to do much independent reporting and frequently post stories that simply reference information published in other sources. Some online sites also allow contributed columns and staff-reported news and it can sometimes be difficult to differentiate one from another. For the Westside Subway Extension, Metro concentrated on well-organized blogs while keeping watch for new sites that developed.

In 2009, Metro launched its own blog, *The Source*, to cover agency news. *The Source* became and continues to be a powerful tool for Metro to control the timing and content of its own messages. Like other blogs, stories on *The Source* sometimes become catalysts for additional coverage by other media. Editor Steve Hymon talks about *The Source* in Chapter 3.

Facebook

In terms of embracing social media, the Westside Subway Extension is perhaps best known for its use of Facebook. In 2008, project consultants suggested setting up a Facebook group to draw people in their twenties and thirties into the planning effort. Although Metro had not yet developed an agency-wide policy for social networking, subway planners and consultants developed written protocols before the Facebook group was launched. The protocols delineated who was responsible for monitoring and posting to the Facebook site, how frequently it would be monitored, what were considered inappropriate comments, and other such policies.

After Metro launched the group, Facebook introduced a way for businesses and organizations to connect through "pages" and this proved more useful than groups for subway outreach purposes. Unlike groups, the new Facebook pages provided greater visibility for the project. Metro posts also appeared in the activity feeds for individual Facebook users when they "liked" or commented on them. Additionally, Facebook provided metrics for the pages which gave Metro more information about those utilizing the social media tool to track and comment on subway planning activities. Facebook's policies did not allow Metro to simply repurpose the subway group as a page. Metro had to create and promote a new Facebook page and users had to "elect in" to the new page by "liking" it, regardless of whether they belonged to the older group. Metro spent many months encouraging group members to make the switch. Staff began decreasing the amount of new information posted to the group and ultimately ceased posting updates entirely.

As of June 2013, the project page (now called Facebook.com/PurpleLineExt) boasted more than 2,900 "likes" and has proven a valuable tool for communicating project planning and related issues. Metro's subway planning staff can post information and news about a variety of topics swiftly and easily without preparing and issuing press releases or waiting for the agency's web staff to post a project update. Web staff updates required half a day or more to accomplish; Facebook updates are almost instantaneous. Furthermore, Facebook is ideal for letting followers know about new project information such as the release of a new fact sheet, a report to the Metro board of directors, or general news about

Figure 4.7 Sample posts from the Westside Subway Extension Facebook page including agency updates and an extended conversation among commenters.

transportation funding and policy developments without clogging their inboxes with multiple e-mail announcements.

Because of social media's interactive nature, individuals interested in the project can talk to one another via the Facebook page. This has led to some extended online conversations. Metro endeavors to post new items to the page every weekday, most commonly links to stories about the subway or other items of interest for those following the project (Figure 4.7).

Just as project opponents and skeptics are welcome to express their opinions to the agency in writing and at public meetings, Metro is also open to including all views about the subway on the Facebook page and posted stories that are skeptical of the project or opposed to it. While other agencies may not follow the same policy, Metro's subway planners felt that it was important to allow all viewpoints to be heard, especially during the long planning process and before project decisions were made by the Metro board.

Since this effort began, Metro developed an agency social media policy and many of its ongoing planning studies, construction projects, and other functions now have their own Facebook pages. Metro's creative services unit developed design standards to ensure that all the pages have a similar look.

Twitter

In 2009, Metro subway planners set up a Twitter account for the Westside Subway Extension project (@WestsideSubway). One of a dozen Metro Twitter accounts covering everything from service alerts to bicycle information, the account had more than 1,000 followers by June 2013 (the account has been renamed @PurpleLineExt.). Unlike Facebook communications, subway-related tweets are not necessarily posted every weekday. They are used primarily for specific purposes. Metro uses Twitter to send out important information about the project, for example, notifications of upcoming meetings, announcements about releases of documents, reminders of approaching comment deadlines, and noting key milestones or project decisions such as approval of the project's final EIS/EIR or the record of decision award. When Metro's own blog, *The Source*, posted stories explaining some of the more detailed concepts behind subway planning, links to these stories were often tweeted as well.

Metro also began live-tweeting its subway community meetings and developed an interesting technique for this purpose. Most meetings included a formal presentation that passed through many rounds of editing and practice before the meetings took place. Metro used these preparations to develop a "Twitter script" before the meeting that broke each part of the presentation into 140 or fewer character tweets. Here are some scripted tweets from a public meeting in October 2012:

@westsidesubway: Tonight we're walking you through what to expect in the next 18 months or so as we prepare to begin construction for phase 1 of the subway

@westsidesubway: Did you know that when the subway is built, anyone will be able to take the line from Pershing Square to UCLA in 25 minutes?

@westsidesubway: Work will start near Wilsh/La Brea this fall & continue through pre-const phase. Metro will notify stakeholders in advance.

@westsidesubway: Special care will be taken due to prevalence of fossils here. Findings given to Page Museum.

During the meeting, the person posting the tweets could cut and paste the parts while focusing mainly on significant deviations and, more importantly, on public comments and questions. Newer services have since become available that allow a pre-planned Twitter script to post at regular intervals when a meeting begins.

Web Streaming and Video
Metro used videos and web streaming to build support and to share information with the community. During the AA phase of the study, Metro produced a video and posted it on the project website. The video is a slightly shorter, recorded version of the presentation given at public meetings during the early scoping period; it is not a recorded and edited version of the meetings. Instead, it shows individuals from the planning team providing information about the AA at this very early stage. As a result, Metro could make the presentation available to the public from the start of the comment period. At a later stage, Metro recorded a board committee presentation about seismic and other safety issues in the vicinity of Century City. Videos showing the presentation and the follow-up press briefing were posted to the project's website and their availability was announced on the project's Facebook page, via Twitter, and on Metro's *The Source* blog.

To further build public support for the project, Metro produced another video titled *A Subway Story* in 2009. As a first step, Metro defined the "takeaway" it hoped to achieve when the lights came up. The goal was to communicate the following: (1) tunneling technology had advanced, (2) Metro had the expertise to build the subway, and (3) the benefits would make the project worth the effort and expense for all of Los Angeles. An approach to the video evolved from this goal, a director was hired, and a script was developed. After script revisions and approval by Metro, the video moved into production and post-production. Metro stayed involved at every step.

Ultimately two versions of the video were created; one was 8 minutes long and one lasted 5 minutes. Metro showed the video at its own meetings and those hosted by others. The video was also promoted through *The Source* and on Facebook. Both versions are still available on YouTube, Metro's main website, and on the project website. See Figure 4.8.

Figure 4.8 Opening scene from Metro-produced *A Subway Story*. (Screen capture from YouTube retrieved from https://www.youtube.com/watch?v=nz95Gr6 VVLM. With permission from Metro.)

Metro also began web-streaming its community meetings for the subway in 2010. Individuals could view the web-stream live during meetings or watch later when the videos were posted online. Metro would note the availability of the video and provide the link on the project's website, Facebook page, and Twitter feed. Each cycle of Metro meetings was typically conducted as a series with identical presentations held within a few days of each other at locations throughout the study area. To save money, only one of the meetings was web-streamed when presentations were identical. However, Metro made a point to move the locations of the web-streamed meeting as the project progressed.

Incorporating Comments from New Media
Comments posted to online news stories, blogs, and Facebook informed the planning process, and the use of social networking was documented in the project's EIS/EIR. However, Metro did not include these online comments in the compilation of official comments for the project and treated them like letters to an editor. As noted previously, Metro encouraged online commenters and Facebook fans to make their views known at meetings or in writing via U.S. mail, e-mail, fax, or hand-delivered letter,

particularly during formal comment periods such as project scoping and the release of the draft EIS/EIR.

Metro did not include Facebook comments in the formal record during these periods primarily because it could not guarantee that all such comments would be captured. Throughout the effort, a team member monitored the page regularly and took screen shots to categorize and tally the postings. Newer services that were not available when subway planning began would now allow Facebook postings to be fully captured as official comments in future studies as long as Metro could ensure that all such postings were captured in a way they could be maintained and organized. Metro also had to determine what to do with the online discussions among page members and whether making such comments official would limit the agency's ability to remove inappropriate remarks.

Agency Protocols
Metro developed a set of protocols and procedures to ensure that social media messaging was consistent and effective.

Messengers
From the very beginning, Metro ensured that only a handful of individuals were authorized to speak publicly for the project: the project director, community relations manager, media relations personnel, and key agency executives. From time to time, media queries were directed to others within Metro who had specialized knowledge about finance, project construction, or other specific issues. These questions would normally go to designated messengers who would ask an appropriate specialist to respond to the query. Of course, members of the Metro board of directors could and would speak publicly about the project as they saw fit.

This group of key messengers had sufficient seniority and awareness of the breadth of agency activities to ensure that messages were conveyed well, fit within overall project goals, and were consistent for the project and agency as a whole. The limited number of messengers also ensured that a key group of individuals could come together quickly to address rapidly evolving issues as necessary.

Message Consistency and Content
Regardless of the medium, Metro ensured that messages were consistent even though each form of communication required different practices and standards. A report posted on a website can provide greater depth of information than a 140-character tweet, but each form of communication

has its place. It is critical to ensure that messages are consistent and complement one another regardless of how they are delivered and who their audiences are. Inconsistencies will be discovered and highlighted by project critics and others, and will undermine credibility for the project and the agency.

Simplifying Messages

Many social media forms require that messages be simplified. Metro limited in-depth explanations and technical jargon to reports. A Facebook post, tweet, or blog post has to be more synthesized. Metro learned that it is vital to clearly identify the key points it wanted its audience to get from technical reports, articles, and other project information. Planners attempted to highlight those points with direct and compelling language when using social media.

Message Frequency

Social media is, in fact, social. Metro worked to ensure that its social media posts remained fresh through frequent updates. It may take weeks or months to develop a report and a day or two for staff to get it posted on the web. Whenever new information was posted to the project website, Metro would link to it from the Facebook page, tweet the link to project Twitter followers, and ensure that a story was posted to *The Source*. At other times, project staff endeavored to use social media in other ways to keep people engaged. As noted earlier, new links were posted to the project Facebook page virtually every weekday. These included progress reports, updates about study staff activities, and links to interesting stories about the project that appeared elsewhere. At times, staff posted links to stories and information about other projects or issues that were relevant to the Westside Subway Extension, such as other Measure R projects, New York's Second Avenue Subway, San Francisco's Central Subway, and federal transportation reauthorization.

Critical Messages and Inappropriate Comments

Due to the very interactive nature of social media, the biggest concern for public agencies is often whether to allow critical messages and how to handle vulgar or otherwise inappropriate comments. Metro recognized that social media was different from the agency's own website where it could fully control the content. Metro welcomed all sorts of messages on its Facebook page, made a point to post stories with different views, and allowed posting of negative comments, even during the planning stages of the project. This policy also proved to be important for strengthening Metro's credibility. Project fans and skeptics could see that all viewpoints were welcome and project outcomes had not been determined before the appropriate stage in the environmental analysis process.

Before launching its Facebook presence, Metro was especially concerned about posts using vulgar or inappropriate language. While these have had to be addressed from time to time, incidents have been minor. Users seem to understand what is acceptable. Over the first two years, only a handful of posts were removed for these reasons. In later planning stages, when particularly controversial issues were in focus, heated language would at times bubble over onto the Facebook page. Metro's response was always the same. The message with inappropriate language was removed, and a response was posted reminding everyone that their interest and passion in the project was welcome but that language had to be appropriate for all audiences. Any individual who initially posted an inappropriate comment was then invited to repost his or her views consistent with those provisions. Metro has now developed comment guidelines that are posted on all the agency Facebook pages.

Responding to Comments and Questions

Whether on external blogs or its own Facebook page, Metro learned to be patient about responding to comments and questions. As interest in the subway grew, members of the public often responded to comments and questions about the project that were posted on outside blogs or on the subway Facebook page, sometimes providing links back to Metro's material to demonstrate their points. These posts were viewed in some sense as being more independent than responses that came from Metro. Metro provided responses online if none came from the public within a reasonable time, if further clarification was needed, or occasionally to confirm the veracity of an independent post.

Lessons Learned

Metro took two lessons from this effort. The first is that utilizing social media can augment study outreach but does not fully replace more traditional methods. It is still necessary, for instance, to use e-mail, U.S. mail, hand delivery of notices, and paid display ads in community papers to promote public meetings. Metro's experience is that well targeted U.S. mail and hand delivery increased meeting attendance and expanded the project's database. Display ads did not visibly produce greater turnout but were important for helping to establish project identity and also for targeting ethnic communities.

On the other hand, some of the newer strategies seem to appeal to those who may not want to or are unable to attend meetings. Social media allows participation in planning efforts by those who work in the study area and live elsewhere and are unavailable to attend weeknight meetings. New media methods certainly seem to have succeeded in drawing in younger participants. According to the Facebook Insights metrics, nearly 36% of those who "like" the subway on this social media platform are between 25 and 34 years of age. Although no data reveal the ages of meeting attendees or those sending in comments, observation of the study team indicates that social media participants appear to be significantly younger than those who attend project meetings. It is also likely that utilization of social media will grow among older age groups as they increase their use of social media and as younger users age.

The second key concept is that social media strategies must be well thought out and fully integrated into the communications effort for a project. At every stage, Metro endeavored to have a very clear idea of what audiences it wanted to reach, what it wanted to communicate, and the role that each form of communication played in achieving those goals. Evaluating whether, what forms, and how to best use social media is no different from evaluating these aspects of traditional media. Any communication vehicle will only be as effective as the overall outreach program and cannot alone compensate for an unclear message, poor understanding of audiences, or the lack of an overall communications goal.

Austin Strategic Mobility Plan

Austin is a fast growing region that has felt its fair share of growing pains. Traffic congestion had grown so substantially that many in the business community argued that congestion was the number one barrier to further economic growth. Looking for solutions to the gridlock, the City of Austin undertook its Austin Strategic Mobility Plan (ASMP) with the hope for a successful bond referendum that could lead to multimodal solutions. The ASMP* focused on providing solutions that could quickly begin to address congestion. The plan was guided by values of sustainable growth, safety, regional coordination, mobility choices, economic development, environmental stewardship, and neighborhood connectivity—everything

---

* http://www.austin-mobility.com/

from walking trails and bike lanes to urban rail and roadway projects. A successful November 2010 Mobility Bond package provided the first round of funding.

In 2009, as the city prepared to begin ASMP, there was a desire to include new media in the participation process. The city, in partnership with the Capital Area Metropolitan Planning Organization (Capital Metro) and the Texas Citizens Fund, applied for and received a $98,000 FTA Public Transportation Participation Pilot Program grant to experiment with social media in the planning process. The Social Networking and Planning Project (SNAPP) piloted, tracked, and evaluated the use of social networking to build relationships and gain valuable public participation. New media represented just one piece of a larger participatory strategy.

Community Participation Strategies

Austin is famous for technology and sometimes is known as the Silicon Hills. With a relatively young and technology-oriented population, Austin often experimented with the use of new media in its planning projects (Goodspeed 2010). Planners for the ASMP believed that experimenting with new media would allow greater involvement among residents too busy to attend public meetings. The receipt of the FTA grant allowed for experimentation with new media at a level beyond what was possible in previous projects.

SNAPP used many tools that have become commonplace for project planners and added new methods for collection and analysis of participatory data. The ASMP included media outreach, public meetings, and a variety of other traditional tools. This section focuses specifically on the new media aspects of the project.

Project Website

The project website has continued to grow and expand since the start of the ASMP in 2009. Austin-mobility.com indicates how to become engaged in the process, reports progress on urban rail, and provides lots of information on development and expansion of various modes of transportation (Figure 4.9). Members of the public can request an educational meeting in their neighborhood or participate in meetings already scheduled near their homes.

SNAPP had its own website (http://www.snappatx.org) to document the project. A live feed updated every 15 minutes contained the latest social media posts from all sources about transportation in Austin (Figure 4.10).

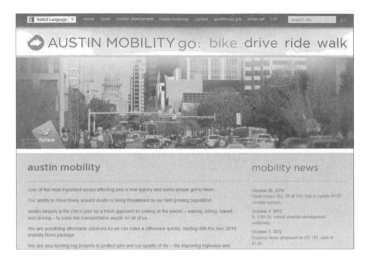

Figure 4.9 Austin Mobility website provides resources to explain the Austin Strategic Mobility Plan and its implementation. (Screen capture from Austin Mobility website.)

New and Social Media Strategies

A 2009 survey found Austin among the top 10 cities using Twitter in the U.S., based on user data indicating their locations (New Media Lab 2009). Based on the significant level of existing social network use, Austin was well positioned to experiment with Twitter and other social media platforms. SNAPP organized a team of facilitators to engage the public via multiple social media platforms including a blog, Twitter, and Facebook.

Blog

SNAPP developed its own blog[*] centered around simplifying complex issues in the ASMP and related plans into plain English. To encourage readership, SNAPP tweeted captivating headlines such as "It's the Parking, Stupid: One Transportation Consultant's Tough Love Approach" (see Figure 4.11). The blogs encouraged some people to provide comments. For example, in a post about car ownership among young people, a reader responded "I think this also discounts the way that technology is changing transportation. One of the reasons I like buses now is that I can sit

---

[*] http://blog.snappatx.org/

Figure 4.10 SNAPP website feed provided a constant stream of social media posts about transportation in Austin. The feed captured a mix of relevant and non-relevant tweets. SNAPP analysts reviewed the tweets database to code for relevance. (Screen capture from SNAPP website.)

and work. If I take the bus it might take 1 hour each way, but I can work the entire time. And I can negotiate that hour into my work day with my employer since I'm productive. If I take a car, it's 30 minutes each way that's just gone. I think the ability to make a commute part of the work day rather than a waste of time is a key reason for moving away from cars" (SNAPPatx 2010). These blog comments were captured in SNAPPatx's database, summarized, and provided to the city in regular reports.

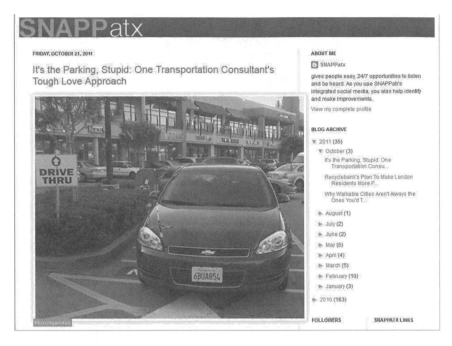

Figure 4.11 SNAPPatx blog focused on explaining key transportation issues and inviting commentary. (Screen capture from SNAPP blog.)

In addition, fostering relationships with bloggers active in transportation was critical. For example, the story of SNAPP's launch in 2010 was picked up by *Capital MetroBlog*, the official blog of the Capital Metropolitan Transit Authority. *The Austinist*, an all-things-Austin website, also picked up the story. These outlets reach a new media-savvy audience and those reading these blogs are people Austin hoped would be active voters interested in supporting transportation improvements. Traditional news media also published stories about SNAPP; for example the *Austin Chronicle* shared news about the launch.

Facebook
SNAPP used Facebook as part of the project. Facebook is an example of a closed network (individual users need to "like" or "friend" to view information posted regularly on the SNAPPatx Facebook page). In addition, searching Facebook pulls back results only from users who have been friended. This creates a challenge for public agencies to find people who

are interested in transportation issues or reach out directly via Facebook (Evans-Cowley and Hollander 2010). The ASMP process had a short time frame with the goal of the SNAPPatx experiment to occur over the course of a few months. This made it difficult to build a robust follower base, although more than 200 people liked SNAPPatx.* SNAPPatx principally used Facebook to share information. The average of 2.3 comments per Facebook post indicates engagement and that SNAPPatx was able to encourage participation. The comments captured on Facebook were included in the database provided to the city.

Twitter
Twitter was a critical platform for SNAPP. Twitter is an open network and most users allow others to view their tweets. Searches can reveal users who are writing about transportation issues more easily than users can be found on the closed-network Facebook. Of the relevant posts from the blog, Facebook, and Twitter, the vast majority (99.7%) came from tweets. All posts from all platforms were continuously streamed on the SNAPPatx website, allowing visitors to see the transportation topics being discussed.

SNAPP realized that by writing an algorithm that would scrape Twitter's feed it could capture a large volume of tweets about transportation in Austin. The algorithm looked for combinations of keywords such as "atx" or "Austin" to indicate location and "bus" to find people who wrote about bus service. The algorithm was successful in capturing almost 50,000 tweets (Evans-Cowley and Griffin 2012). A small percentage came from SNAPP; over the course of the project, SNAPP sent out more than 1,000 tweets encouraging participation. The remainder came from members of the public who tweeted about transportation issues. The tweets were then reviewed for relevance. For example, tweets where "Austin" referred to a person rather than a city were excluded. Trained facilitators constantly reviewed the tweets as they came in. When a relevant tweet was identified, the facilitator responded to the person tweeting about a transportation topic. Below is an example of the dialogue between Twitter user @daVinci70 and @SNAPPatx.

@daVinci70: is sitting in downtown Austin waiting for her next bus. Just people watching and enjoying the casual conversation of people passing by.

---

* http://www.facebook.com/pages/snappATX

@SNAPPatx: @daVinci70 That sounds like a nice way to pass the time waiting for a bus. Did it arrive on time? How was the ride? #snappatx

@daVinci70: @SNAPPatx yes, the bus arrived on time. However, the driver was accelerating quite recklessly. Is he a retired race car driver?

@SNAPPatx: @daVinci70 Well Formula 1 fever is afoot in ATX. Are you a regular bus rider? Was this an unusual experience? Did you feel unsafe? #snappatx

@daVinci70: @SNAPPatx yes, I did feel very unsafe. And I am not a timid person but his rate of acceleration caused violent jerks when he had to stop.

Through this dialogue the SNAPP facilitator was able to gain information about the timeliness of the bus service and safety. Austin officials were interested to know about the individual users and whether they were Austin citizens. As with any online network, people can be whoever they want to be and reveal a little or a lot of information. For example, @daVinci70's profile picture reveals that she appears to be an African American woman. Her profile description states, "A self proclaimed bibliophile, cinephile, proud mother, and diligent pursuer of knowledge, truth, and justice for myself and others." Her location is indicated to be Austin, Texas.

Public officials raised some concern that people may purposefully misrepresent themselves online. The unique aspect of this project is that the SNAPP facilitators sought people already talking about transportation rather than people approaching the city directly to talk about transportation. It is likely that @daVinci70 represented herself honestly.

Every week all the collected tweets were categorized by theme and topic. The theme related to strategic mobility goals such as regional integration, safety, and economic development. Then the topics were determined based on specific tweet content. "Austin really needs to invest in four-lane highways. Maybe even five. Anything … anything that's more than three!" would fall into the investment and economic development theme and the topics are highways and automobiles. The team would then determine whether a positive or negative sentiment was expressed. For example, the "I hate Austin traffic. #getittogether people!" tweet expresses a negative sentiment.

The city received regular reports that provided a breakdown of positive and negative sentiments for each transportation topic discussed. For example, not surprisingly, significant negative sentiment was expressed when traffic was discussed. The city staff found it helpful to gain a simple understanding of the sentiments expressed.

Project Protocols

Project planners developed a set of procedures and protocols to ensure consistency in project communications.

Messengers

Significant conflict about messaging arose between the city and the Texas Citizen's Fund. While SNAPP utilized trained facilitators skilled at crafting messages appropriate for social media, the city was concerned that the messages might have been biased. The SNAPP group understood the need to create provocative messaging that would attract attention and encourage interaction. The city lacked the time and expertise to effectively assist with messaging. In future projects, the determination of responsibility for messaging needs careful consideration.

Critical Messages and Inappropriate Comments

The SNAPP project had its share of dissenters who expressed negativity. Their comments were welcomed and SNAPP facilitators probed deeper to find out why participants felt the way they did. Some of the posts included profanity. A filter was set in the algorithm to ensure that posts with profanity were captured and excluded from the feed on the website.

Responding to Comments and Questions

The SNAPP facilitators worked hard to respond within minutes whenever possible to posted messages, even during the wee hours of the night. The timeliness of the responses allowed for participation in real time and enabled participants to feel they were heard.

Lessons Learned

The key take-away from the Austin experiment was the need to develop analytical tools. For example, the city really appreciated the sentiment analysis. During the project, only a simple positive or negative sentiment analysis was used. The city staff really wanted to know the stories behind the sentiments. More detailed sentiment studies were undertaken using Linguistic Inquiry and Word Count software. The more detailed analysis created a deeper understanding. For example, when people discuss traffic, they are doing so in present time and space—they are likely tweeting while stuck in traffic.

Another key lesson was the need for trained facilitators to engage in dialogues of 140 or fewer characters. Special skill is needed to find ways to be concise and quickly get to the point with each communication. SNAPP also learned that peak Twitter activity occurs around 10 p.m. and facilitators were required to work evening hours.

The practice of scraping Twitter to find people discussing transportation provided a great way to reach out to people who would not have otherwise become engaged in the planning process. While the dialogues were brief, they provided a new and different way to allow the public to participate. For the most part, people were happy to participate and those contacted by facilitators via Twitter after they sent transportation-related tweets responded 42% of the time. This is considerably higher than would be expected from a survey for example. Deploying this type of participatory technique requires significant planning and substantial time. SNAPP underestimated the amount of time required to categorize and analyze the tweets.

Findings and Conclusions

Planners in Los Angeles and Austin used social media to connect with citizens during long-range planning projects. For Metro, two key lessons emerged from the efforts. First, using social media can expand outreach but cannot fully replace traditional media methods. While social media appeared to reach those who might not otherwise attend a meeting (especially appealing to young adults), the agency still used e-mail, U.S. mail, hand delivery of notices, and advertising to promote public meetings. Despite active conversations in online social networks, Metro did not incorporate social media comments into the public record for the Westside Subway Extension project.

Second, social media strategies must be well thought out and fully integrated into the communications effort for a project. At every stage of the public planning and review process, Metro worked to hone its messages to ensure its information was clear and relevant to the targeted audience. Any communication vehicle will only be as effective as the overall outreach program.

In Austin, officials used social media as part of an overall communications strategy to encourage public conversation about transportation issues in the metropolitan area. Tracking social media posts helped find people discussing transportation, and responding to those messages helped engage people in the planning process. Analyzing the information contained in social media posts helped city officials understand how citizens felt about local transportation systems and conditions. The Austin SNAPP project highlighted the need to develop analytical tools to gain a deeper understanding of the information and sentiments contained in social media posts. The project also demonstrated that social media conversations take place around the clock. When considering an engagement strategy, it is important to remember that the public will engage at all hours. Organizations should consider whether and how they intend to respond to comments, whether posted during working hours or late at night.

FROM HURRICANES TO CARMAGEDDON: SOCIAL MEDIA FOR REAL-TIME COMMUNICATIONS
Ned Racine and Susan Bregman

Social media provides agencies with an unparalleled opportunity to share information with their customers—often in real time. Twitter is exceptionally well suited for transmitting up-to-the minute information. Many transportation organizations turn to tweets to inform the public about transit service delays, traffic tie-ups, road closures, and construction detours. Some even provide information like waiting times at the department of motor vehicles. Increasingly, transportation agencies have incorporated social media tools into their emergency communication protocols, and some have set up dedicated Twitter accounts for emergency updates and bulletins. Citizens also turn to social media to share information about travel delays and emergency conditions.

This section discusses the vital role that social media plays in communicating information about real-time conditions. Examples cover updates for traffic and transit conditions, construction management for the Los Angeles "Carmageddon" project, crisis management during Hurricane Sandy, and crowdsourced information. The section concludes with an overview of the potential risks arising from using social media to communicate real-time transportation information and strategies for addressing those concerns.

Real-Time Conditions

Delays, congestion, and service disruptions are unfortunate facts of life for the traveling public, and transportation agencies use several tools for communicating incident details. Take the example of a rail delay caused by a mechanical failure. Before the advent of social media, a transit operator might have shared details with print and electronic media through a press release, posted information about the delay on its website, sent out alerts via e-mail or text messaging, provided updates on electronic station signage, and made announcements on the train or at the station.

According to research conducted by Passenger Focus (2011), a British watchdog group that represents the interests of bus and rail passengers, riders want three key pieces of information when facing an unplanned service disruption:

- Accurate prediction of length of delay
- Reason for delay
- Alternative routes

Also important is transparency. The surveyed rail riders showed healthy skepticism about much of the information they received about delays, especially what many perceived as the generic excuse of "signal failure." Riders wanted to learn about delays as soon as the information was available and they wanted to know the real reasons for delays.

Enter social media. While not specifically addressed in the Passenger Focus survey, social media platforms are well suited to provide the kind of information passengers seek when they encounter unexpected delays. Transportation agencies frequently take advantage of social media's real-time features to communicate time-sensitive information to their constituents and generally favor Twitter for these updates.

The Southeastern Pennsylvania Transportation Authority (SEPTA) providing bus, trolley, subway, and commuter rail services in Philadelphia and the surrounding suburbs implemented this approach. In addition to @SEPTA, the agency's main Twitter feed that distributes overall updates, SEPTA has set up individual Twitter accounts for each of its trolley, subway, and commuter rail lines. Another feed provides information on all the agency's bus routes, although separate accounts are not available for specific lines. By offering 26 separate Twitter accounts, SEPTA gives its riders the option to subscribe only to service alerts that are relevant to their daily travels. Figure 4.12 shows tweets from SEPTA's primary @SEPTA Twitter account.

The Pennsylvania Department of Transportation (PennDOT) and the state's 511PA traveler information system (www.511pa.com/) also manage multiple Twitter accounts that provide advisories and updates on road

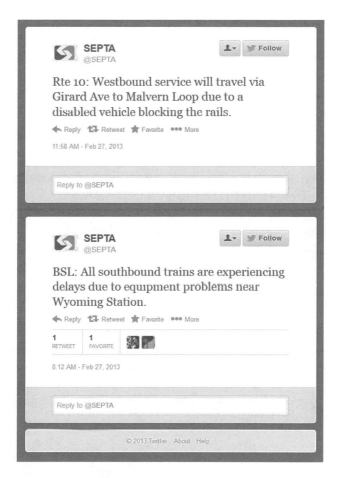

Figure 4.12 The Southeastern Pennsylvania Transportation Authority (SEPTA) uses Twitter to provide updates about service conditions. (Screen capture from SEPTA Twitter account.)

conditions for eight regions along with a statewide feed (@511PAStatewide). Typical tweets are succinct:

@511PAStatewide: Road maintenance operations: PA Turnpike. Turnpike I-76 east. East Exit 39–PA 8, left lane.

@511PAStatewide: Roadwork: I-70 eastbound. East Exit 25–PA 519.

@511PAStatewide: CLEARED roadwork: I-79 northbound North Exit 43– PA 519.

As these examples indicate, each tweet follows a template: type of incident, road, direction, and location. The posts provide drivers with essential details about roadway incidents and, equally important, let them know when normal conditions have resumed.

Social media can also take the uncertainty out of transportation-related transactions like applying for a driver's license or registering a car. Anyone who has visited a motor vehicle department knows the frustration of waiting in long lines. The Massachusetts Registry of Motor Vehicles[*] is one of several state DMVs that post current waiting times on their websites so citizens can check before leaving home or work. South Dakota has taken this concept into the social space and tweets the waiting times at DMV branches (@SDDriverLicense). For example: "Wait time is now 50 minutes at SF Westport and 5 minutes at SF Admin. No waiting in Aberdeen and RC."[†]

The California Division of Motor Vehicles has an active social media presence. It uses Twitter (@CA_DMV), Facebook (www.facebook.com/CADMV), YouTube (www.youtube.com/californiadmv), and a blog (cadmv.wordpress.com). While DMV does not use social media to update waiting times at branches (mobile applications are available to provide this information), the agency encourages customers to "avoid waiting on hold and save time by tweeting @CA_DMV to receive timely responses to customer questions and concerns" (CADMV 2010). DMV tested the impacts of engaging its customers on Twitter. The DMV Blog describes the changes since the pilot started:

> Through the pilot, the department began to respond to customers on Twitter and document all activities to gauge the beneficial value of using Twitter as a means of two-way communication and customer service. The pilot allowed the department to become more transparent and receptive to the online community, removing communication barriers with social network users and broadening opportunities to change consumer sentiment toward DMV.

In July 2010, DMV had 1,675 Twitter followers. By March 2011, the number had grown to 2,803. Measures of engagement increased over this period as well. For example, monthly retweets increased from 79 to 283 and mentions grew from 504 to 965 (CADMV 2011).

---

[*] www.massdot.state.ma.us/rmv/
[†] SF = Sioux Falls; RC = Rapid City.

Construction Management and Carmageddon

Social media has also proven effective in communications with citizens during planned disruptions. In July 2011, a section of the I-405 freeway in Southern California was closed for a weekend to demolish a bridge as part of a highway improvement project. The outreach to prepare people for the closure earned accolades for the project team, contractors, and the network of public safety organizations participating in the planned 53-hour event. Named *Carmageddon* by some unknown wit, news of the closure reached travelers around the world and gave Los Angeles County a boost in civic pride. It showed that drivers would respond to a "we're all in this together" plea and underscored the power of frank and persistent social media.

The San Diego Freeway, commonly called "the 405" by Southern California residents, runs from the city of Irvine in Orange County to the north San Fernando Valley communities (that have the same population as Chicago). Basically shadowing the Southern California coastline, the Los Angeles County section of the freeway passes several institutions of local, regional, and national import: Los Angeles International Airport, Getty Art Museum, University of California, Los Angeles, and the Skirball Cultural Center.

The most heavily used section of I-405 runs through the Sepulveda Pass in the Santa Monica Mountains. The California Department of Transportation (Caltrans), the ultimate owner of I-405, estimated that 100 million vehicles used the Sepulveda Pass section of the freeway in 2010. So while I-405 is often reviled, it remains impossible to ignore. In part, this is because the Pacific Ocean and the Santa Monica Mountains squeeze the Los Angeles Basin, offering drivers topographical obstacles similar to those found in San Francisco or New York City. The area has no high-volume alternative to I-405. Figure 4.13 shows it on a typical day.

As a result, media concern about absolute gridlock began before the official announcement of the July 2011 closure/demolition and grew— which helped to build a social media audience for the closure. One additional factor aided the social media campaign. The July 17–19 weekend would offer unique sights. The mighty I-405 would present ten empty lanes and a landmark bridge would be pounded into concrete shards.

143

Figure 4.13 Typical Interstate 405 traffic traveling through Sepulveda Pass. The photo looks north toward the San Fernando Valley with Sepulveda Boulevard to the east of the freeway. (Courtesy of Metro® 2013 LACMTA. With permission.)

Heart Surgery during a Marathon

The final EIS/EIR approved in January 2008 for the I-405 Sepulveda Pass widening delineated a project to add a 10-mile-long high-occupancy vehicle (HOV) lane from the Santa Monica Freeway (I-10) to the Ventura Freeway (U.S. 101). When combined with existing assets, the HOV lanes would be 72 miles long, creating the longest HOV segment in the U.S.

Before these benefits would arrive, however, the project required realignment of 27 on- and off-ramps, widening of 13 underpasses, and construction of 18 miles of walls. The new lane would be added along the east edge of the highway and dozens of utilities would have to be moved farther east, away from the new freeway lane. A section of heavily traveled Sepulveda Boulevard running alongside the freeway would have to be moved more than 20 feet east as well. As Michael Barbour, project director, later told a reporter, "It would be like performing heart surgery on a marathon runner during a race."

Three bridges would have to be demolished and replaced to span the widened freeway. In a nod to the traffic they carry, only one side of each bridge would be demolished at one time. Moving north, they were:

- Sunset Boulevard Bridge
- Skirball Center Drive Bridge
- Mulholland Drive Bridge

The Mulholland Drive Bridge would be the greatest challenge. Because of its height and design, the 579-foot-long bridge could not be demolished in sections over a series of nights, as could the Sunset Boulevard Bridge. The Mulholland Drive Bridge would have to be demolished one entire side at a time, south side first. For safety reasons, the nation's most traveled freeway under Mulholland Drive Bridge would have to be closed while seven hoe rams shattered the 1960 bridge into nine million tons of concrete debris. This would be the first high-profile challenge for the project's social media channels.

The construction relations team had less than eight weeks to change the behavior of Southern California drivers. Caltrans asked the Sepulveda Pass Widening Project for a 35% decrease in traffic on the adjoining sections of I-405 during the July 17–19 weekend. Caltrans doubted that it could be done. Although Sepulveda Boulevard would remain open as a detour route during that weekend, Caltrans worried that north and south traffic used to filling I-405 would flood I-110 through downtown Los Angeles or I-605 farther east. For the Carmageddon campaign to succeed, drivers would not only have to avoid the Sepulveda Pass, many of them would have to stay home, away from *any* freeway. No series of meetings could reach such a number of drivers; they would have to be persuaded online.

New Commitment to Social Media
Caltrans managed the I-405 project in conjunction with the Los Angeles County Metropolitan Transportation Authority (Metro). The authority took the lead in the project's social media presence. Fortunately Metro's view of social media had been evolving. Rather than taking the "if we don't take part in the conversation, people won't criticize us" stance, Metro's communications department embraced social media. With its official Twitter account, the agency began devoting resources to conversations with stakeholders, both pro and con. Still, Metro had never had an ongoing social media presence for a construction project. Several factors shaped the decision to begin one:

145

- Densities of neighborhoods served by I-405 on Los Angeles' Westside
- Computer literacy of affected local population
- Rank of the Santa Monica/Culver City area as the second greatest job generator in Los Angeles County
- Long-term effect of project on local traffic

Metro created a Facebook page and Twitter account for the I-405 project, but needed someone to be responsible for the mammoth project's social media voice. Once the project's new media officer was in place, the I-405 social media profile rose. Before the arrival of the media officer, the project sent only a few "crucial" weekday closure notices to Twitter. With his arrival, each closure was sent to Twitter and the project website and soon after to Facebook.

A local official recommended the Nixle electronic notification service already popular with California law enforcement and emergency services. Using Nixle as a gateway, postings were written once and pushed by Nixle to the other channels. Frequency grew, and instead of sending closures the usual 5 days a week, the closures were sent on weekends as well. Eventually, closure items were sent three times a day: Day shift closures were sent the night before, night shift closures were sent in the afternoon, and adjustments to the night closures were sent in the early evening. While most postings were written and sent by the construction relations administrator and the new media officer, each member of the construction relations team was given the tools to post.

Construction relations staff mirrored the tone of *The Source*, Metro's well-known blog: informed, educational, and sympathetic. Although the conversation included criticism from followers of the project's social media sites (mostly regarding the pace of the project and whether the finished project would provide enough relief), most comments were questions or praises. As a policy, the new media officer did not answer posts laced with profanity but did answer all questions. In respect for users' time, the I-405 social media sites limited postings to news directly from the project or from Caltrans or Metro.

Besides posting closures three times daily, the new media officer also received real-time updates from the community relations officers in the field and the prime contractor's community relations officers during work days. For example, an oil spill on Sepulveda Boulevard resulted in the following morning posting:

TU-SUN: Oil spill on Sepulveda Bl near Getty ramps closes traffic in both directions until noon. Spill not project related.

Often a posting grew from a closure extended from its scheduled pick-up time:

WE-WIL: Closure of Sunset Bl westbound 2 lane extended to 4:30pm.

Some posts covered closures that were cancelled several hours before the work was to begin:

TH-MUL: Closure of Sepulveda Bl at Skirball Center Dr cancelled for tonight.

The mysterious capital letters at the beginning of each posting are codes. They represent the day of the closure (MO, TU, WE, TH, FR, SA, SU) and the segment of the closure: Wilshire, Sunset, and Mulholland (WIL, SUN, MUL). For events exceeding the boundaries of any segment, the ALL designation was used. The code system allowed busy viewers to scan the closures that mattered to them and also provided a breakdown of work by segment on the project website.

Sending these updates became routine and covered 40 to 50 night shift closures, 3 to 5 additions or subtractions to night shift closures, 10 to 15 day shift closures. As of this writing, the I-405 project's construction relations team sent over 27,000 postings to its social media sites since December 2009; they also appeared on the project website. Because of the real-time nature of the updates, quality control rested with the new media officer, who stayed in close touch with the construction managers to ensure accurate content.

Website Matrices Explode

By May 2011, when elected officials and agency leaders announced the July demolition of the Mulholland Drive Bridge, the project already had thousands of followers on its social media channels. It had established a conversation and a tone. As part of the conversation, the project followers, particularly on Facebook, protected the tone of the conversation, occasionally criticizing rambling or aggressive postings. Table 4.1 shows the number of monthly followers for the project's social media accounts.

Although the Nixle subscribers never grew much beyond 346, many of them were public safety agencies such as the Los Angeles Police Department and the Los Angeles County Sheriff's Department. Indicative of the importance of I-405, many of the Facebook and Twitter followers

Table 4.1   Social Media Traffic: Monthly Followers in 2011

Site	May	June	July
Twitter	1,995	2,312	3,670
Facebook	873	1,780	31,750
Nixle	316	346	282

were influencers such as radio and television stations, particularly traffic reporters. Other followers included hotels in the beach cities to the south and businesses in the San Fernando Valley.

Many of Metro's outreach activities directed visitors to the I-405 website (www.metro.net/projects/i-405). In July 2011, as stories about Carmageddon began appearing in news outlets around the world, the website numbers exploded. During the last few days before the Sepulveda Pass closure, the I-405 website constituted 23% of all traffic to Metro's website—by far the most ever for a construction project. Peak readership came on Friday, July 15, 2011, the day before the closure, when the project website attracted 117,795 page views. The interest in Carmageddon drove unprecedented traffic to Metro's other pages as well. From July 1 through 18, the agency's home page at www.metro. net received 956,733 visits and 2.6 million page views, equivalent to an entire month's normal traffic. Perhaps most impressive, on Thursday evening, July 16, "Carmageddon" became the number one search string on Google.

Why Did It Work?
When the demolition date was announced, the outreach plan had four broad goals, each targeting times before and during demolition:

- Conduct outreach and communicate with local community residents.
- Coordinate with first responders and work to ensure emergency plans are in place.
- Communicate and coordinate with institutions and organizations.
- Conduct outreach and communicate with traveling public and commuters, particularly sharing detour options.

Metro's social media outreach for Carmageddon was unprecedented for a freeway construction project. Construction relations made the I-405 website the warehouse for renderings, videos, and maps. Consequently, many social media postings were devoted to announcements of new content arriving on the website and reminders that content could be found there.

Figure 4.14 Leading up to Carmageddon weekend, the Mulholland Drive Bridge Demolition page became the home page of the I-405 project website. (Screen capture from I-405 website. http://www.metro.net/projects/I-405/. With permission.)

To make finding content easier, Metro made the Mulholland Drive Bridge Demolition page the default home page for the I-405 website 2 weeks before the closure (Figure 4.14.) Each week, several postings reminded the social media audience that the demolition and closure were coming. Each reminder addressed a different segment of the closure. For example:

MO-MUL: During July 17–19, Sepulveda Bl will be open for local traffic only.

TU-ALL: On Friday, July 17, freeway ramps will close as early as 7 pm.

MO-MUL: During July 17–19, Sepulveda Bl will be open for local traffic only.

TU-ALL: On Friday, July 17, freeway lanes will close as early as 10 pm.

Determined to generate enough original content for viewers, Metro's web team and construction relations staff developed downloadable ads that webmasters could post on their websites. The gem of the downloadable files was a countdown clock that showed the days, hours, minutes, and seconds until demolition began. Multiple social media postings sent the

149

link for the downloads. The links were also sent via e-blasts to more than 6,500 regional stakeholders and 1,300 e-mail addresses of I-405 followers. By July 20, 200 websites posted the countdown clock and the clock attracted more than three million views.

Although the frequency of posts may appear excessive, looking back, the repetition was required. Some people simply wait until the last minute to focus on a project's details and, despite extensive local and national press coverage, the new media officer continued to receive basic questions as late as the Friday of the closure:

- Will the closure really happen this weekend?
- Will Sepulveda Boulevard be closed as well?
- When do you expect the closure to end?
- Is I-10 closed as well?
- Will emergency vehicles be able to get through the closure?
- Why do you have to close the freeway at all?

The magnitude of Carmageddon combined with construction relations personnel's willingness to welcome help from every corner resulted in unusual acts of support. The new media officer negotiated free advertising from Facebook. As did many social media channels outside Metro, the Facebook ad directed viewers to the I-405 website. Facebook estimated the ad generated 6.6 million impressions. LA Inc, also known as the Los Angeles Tourism and Visitors Bureau, sent more than 400 Twitter postings nine days before the closure and during the closure. One ancillary benefit of Carmageddon was that local elected officials began sharing social media postings, particularly retweeting Twitter updates.

Outreach for Carmageddon earned Metro and Caltrans a Project of the Year award from the California Transportation Foundation, but the key mark of success had always been removing traffic from the Sepulveda Pass that July weekend. Total vehicle miles traveled (VMT), a key Caltrans measurement, dropped 12% across Los Angeles County that weekend. Caltrans estimated that drivers using Southern California freeways spent 8 to 10 fewer minutes completing their trips. Thanks to the extensive outreach, drivers did not simply shift to other routes; they stayed out of their cars. By that measure, the social media campaign succeeded. Figure 4.15 shows street cleaners sweeping a car-free stretch of I-405 on the Saturday of Carmageddon weekend.

By all measures, the Carmageddon campaign was a success. Caltrans was pleased that a history-making gridlock did not occur, Metro was gratified that its outreach efforts won acclaim from elected officials and

Figure 4.15 Street cleaners sweep empty I-405 during Carmageddon weekend. (Courtesy of Metro® 2013 LACMTA. With permission.)

local media. Hundreds of businesses and agencies were happy that they helped avoid a transportation nightmare. Westside residents acted collectively and responsibly. For Metro, Carmageddon showed social media outreach to be a mature tool that could reach a diversified population, and the agency decided to add social media components to all future construction projects.

Emergency and Crisis Communications

Sometimes transportation agencies find themselves in the middle of a crisis. Perhaps a train derails, a series of tornadoes shuts down air traffic, or a flood makes roads impassable. Unexpected events like these may involve a significant public safety threat, ongoing uncertainty, and time urgency (Tinker and Fouse 2009). A well prepared organization will have a plan in place to guide communication with the public throughout these rapidly evolving situations. Forward-thinking organizations have incorporated social networking tools into their emergency communications strategies to "transmit critically important information immediately to as many people as possible" (Tinker and Vaughan 2012).

Local, county, and state governments across the U.S. have incorporated social media into their emergency preparedness and communications plans, and transportation organizations are following suit. After Hurricanes Katrina and Rita battered the Gulf Coast in 2005, the Mississippi Department of Transportation (MDOT) updated its emergency communications strategies (Volpe 2010). As part of its overall planning efforts, the department set up route-specific Twitter accounts to provide storm-related information for the state's six main evacuation routes. MDOT's goal for setting up the accounts was to provide real-time updates on travel conditions, contraflow routes, fuel availability, and roadway openings. The agency first had a chance to test the accounts when Hurricane Isaac struck in 2012. MDOT tweets during that storm included the following:

@mdot_i59: MS south of I-20 is under flood watch through Wednesday. #Isaac to bring heavy rains and flash flooding. Motorists urged to drive safely.

@mdot_us49: All lanes Hwy 90 from Bay St. Louis Bridge to Biloxi Bay Bridge are closed due to #Isaac flooding.

@mdot_us98: MDOT crews and others are working hard to restore MS roadways after #Isaac. Slow down in work areas, please!

@mdot_i59: Hwy 43 Spur at North Beech Street in Picayune is now OPEN to traffic.

In October 2012, Hurricane Sandy provided an opportunity to observe how transportation agencies use social media during a weather emergency. This historic storm devastated communities along the eastern seaboard and created massive damage for the New York City subway system. Multiple transportation agencies suspended transit service or closed roadways, bridges, and tunnels during the storm to protect people and property. The New York Metropolitan Transportation Authority (MTA) and New Jersey Transit would be unable to restore full service for months.

When the MTA made the decision to shut down bus and rail service in anticipation of the storm, officials used a combination of traditional and social media to spread the word. The agency issued a press release that was posted on its website and also used Twitter to notify the public. The authority issued the following tweet an hour before service was scheduled to shut down:

@MTAInsider: Suspension of service on the NYC Subway, LIRR & Metro-North will begin in one hour. Start leaving for your destination soon. #Sandy

This tweet was issued after the service suspension was complete:

@NYCTSubwayScoop: #ServAdv: The orderly shut down of the subway is complete. You may see diesel or work trains out there but passenger service is suspended.

The MTA also oversees bridges and tunnels in New York City and used Facebook to let people know about hurricane-related closures and vehicle restrictions.* One example:

> Due to high winds certain types of vehicles are not permitted to cross the Verrazano-Narrows, Throgs Neck, Bronx-Whitestone, Cross Bay Veterans Memorial, Marine Parkway-Gil Hodges Memorial, and Robert F. Kennedy bridges at this time. These include tractor trailers, step vans, house trailers, cars pulling trailers, mini buses, motorcycles and vehicles with open backs carrying plate glass, or any cargo that may become dislodged.

Throughout the storm and the clean-up efforts, MTA communications staff used Twitter and Facebook to keep the public updated about the

---

* https://www.facebook.com/pages/MTA-Bridges-and-Tunnels/244246159423

Figure 4.16 MTA used Flickr to post photos of damage from Hurricane Sandy including this boat washed up on the tracks of the Metro-North Railroad near Ossining, New York. (From: Metropolitan Transportation Authority via Flickr. Licensed under Creative Commons Attribution 2.0 Generic (CC by 2.0). With permission.)

status of the system and shared storm-related photos and videos with the public and the media on its Flickr account.* These images were widely shared among media outlets and provided a compelling account of the hurricane and its aftermath. Figure 4.16 shows a boat that washed up on the tracks of the Metro-North Railroad near Ossining, New York.

Departments of transportation also used the social space to alert motorists about road conditions. While the hurricane caused water-related damage for coastal communities, the storm brought snow to the mountains of West Virginia. The West Virginia Department of Transportation used Twitter[†] and Facebook[‡] to provide travel updates. Some sample Twitter posts are presented here:

@WVDOT: As of now, looks like biggest weather threat will not be until Monday. Things could change. #sandy #frankenstorm

---

* http://www.flickr.com/photos/mtaphotos/
† https://twitter.com/WVDOT
‡ https://www.facebook.com/WVDOT

@WVDOT: Popular item in carts today: wiper blades. Good idea, with heavy rains & snow in forecast.

@WVDOT: Whiteout conditions are possible in areas under Blizzard Warning. Take it slow or just stay home!

@WVDOT: Tucker County: WV-32 is blocked just south of Davis, near the Tuscan Ridge Community, due to a two vehicle accident. Flagging traffic.

@WVDOT: Accidents on Interstates: I-79, I-77, I-64. Snow is still coming down. Some roads may appear clear, but are icy. Stay home! #WVsandy

Airports, too, were affected by Hurricane Sandy and the BWI-Thurgood Marshall Airport outside Baltimore was among those using social media to provide storm-related information to travelers. BWI posted this statement on Facebook after the hurricane:

> BWI Marshall Airport withstood the historic coastal storm yesterday and last night without significant damage or flooding. A big THANK YOU to airport, airline, TSA, and concessions employees for your good work during the massive storm.
>
> Very limited airline service is resuming this afternoon at BWI Marshall. Most flights remain cancelled.
>
> Due to the scope of the storm, airlines will continue working to reposition aircraft and resume operations in coming days at BWI Marshall and at airports throughout the eastern U.S.

A few hours later, BWI welcomed the first flight to land after the storm:

> The first post-Sandy passenger flight that arrived at BWI Marshall Airport this morning was a big one!

The accompanying photograph showed a jumbo jet on its final approach (Figure 4.17).

Even bicycle-sharing programs saw the impact of the storm. Boston's Hubway used Twitter* to keep its riders apprised of changing weather conditions. The program posted updates via Twitter before, during, and after the hurricane:

@Hubway: Due to Hurricane Sandy, Hubway will be closed temporarily from 10 am Mon, Oct. 29 until conditions improve. Stay tuned, stay safe. #PleaseRT

---

* https://twitter.com/hubway

Figure 4.17 BWI-Thurgood Marshall Airport used Facebook to thank its workers and welcome the first post-Sandy flight. (Screen capture from BWI Marshall Facebook page.)

@Hubway: See any #SandyDamage to our Hubway stations? Please let us know, and thanks! #BuddySystem

@Hubway: We expect to be back online for the afternoon commute barring any surprises.

@Hubway: UPDATE: Hubway stations reopened as of 1:30 pm, be sure to use caution on the roads while crews finish cleanup work. #SafetyFirst #ByeSandy

Crowdsourcing Real-Time Information

Not every transit agency or DOT provides real-time information, and sometimes drivers and transit passengers fill the information gap by sharing notes from the field. Transit passengers often turn to social media, especially Twitter, to share their frustration when trains and buses are late, and some developers have found a way to capitalize on this fact of commuting life. The Clever Commute* is an e-mail and Twitter-based service by which transit riders share real-time information about travel conditions with their peers. The CTA Station Watch† website bills itself as a crowd-sourced information stream about the progress of capital construction projects at nine stations on the Chicago Transit Authority (CTA) Red Line. The site administrators set up a Twitter account, Facebook page, and Flickr group and encouraged readers to share news, photographs, and construction updates for these stations.

Some third-party developers and government agencies chose instead to integrate and share real-time information via mobile applications for smartphones and tablets. While this book does not focus on mobile applications, several are worth brief mentions:

- Tiramisu (discussed in Chapter 5) provides a mobile platform for passengers on the Pittsburgh-based Port Authority of Allegheny County's buses to share information about bus location and levels of crowding (www.tiramisutransit.com).
- The Transportation Security Administration released a mobile application called MyTSA that provides crowdsourced information about waiting times at airport security checkpoints along with other resources related to travel safety (www.tsa.gov).
- Waze is a community-based mobile application that encourages drivers to share information with one another about traffic hot-spots, construction delays, accidents, and speed traps (www. waze.com). Drivers can participate in two ways. Users can drive while the application is open on their phones, and Waze will automatically collect traffic and other road data. Drivers can also share information more actively by reporting accidents, detours, and the like. (For safety, the application disables typing while the vehicle is moving.)

Mobile applications and social media can work hand in hand, and it is not unusual for real-time information about an agency to be available on multiple platforms. This has been the case for transit agencies—especially major urban operators. An active developer community may be available to create free and paid mobile applications. Applications like Tiramisu and Waze that use crowdsourced information are particularly valuable in situations where transportation agencies cannot or do not provide

---

* http://clevercommute.com
† http://ctastationwatch.com/

real-time information. For many public agencies, however, social media provides a free and readily accessible platform for communicating time-sensitive information to citizens and the media. Agencies should keep in mind that a social media platform may reach a wider audience than their own independent sites. Agency-specific applications work better when information should be individualized (e.g., service alerts) or when the functionality required (e.g., a location-aware feature) goes beyond that available on typical social media platforms.

Questions and Concerns

While social media provide a platform for agencies to communicate with citizens about planned and unplanned transportation disruptions, they present some potential drawbacks. As the widespread use of mobile devices has empowered citizens to document and share information during times of crisis, the potential for rumors and inaccurate information has grown as well. Despite such worries, Lindsay (2011) reports that some researchers determined that the "information gleaned from social media is generally accurate, suggesting that reports about the spread of misinformation during incidents may have been exaggerated."

Another concern is use of social media to intentionally post malicious or misleading information during emergencies. During Hurricane Sandy, while public safety agencies and media outlets posted official storm information on social media platforms, an individual calling himself @ComfortablySmug posted false updates on Twitter. For example, he posted this misleading information:

@ComfortablySmug: BREAKING: Con Edison has begun shutting down ALL power in Manhattan.

His posts were widely retweeted until official sources began rebutting the claims. The media subsequently discerned his identity and @ComfortablySmug stopped tweeting shortly thereafter (Stuef 2012).

Finally, using social media, and especially Twitter, to update motorists about real-time road conditions raises an obvious safety concern. Oklahoma DOT (@OKDOT) and Arizona DOT (@ArizonaDOT) are among the state departments of transportation that inserted disclaimers directly on their Twitter pages to remind followers not to text or tweet while driving.

Potential drawbacks like these should not discourage agencies from using social media for real-time communications. For most organizations, the many benefits of using social media outweigh these concerns. Agencies should make fully informed decisions and take steps to avoid the pitfalls associated with social media use for emergency and real-time communications.

Conclusions

Many transportation organizations use social media to inform the public about transit service delays, traffic tie-ups, road closures, and construction detours. As they become more comfortable with these communication tools, public agencies have begun to incorporate social media into their emergency communication protocols as well. Citizens also use social media to share information with fellow travelers about delays and emergency conditions.

This section discussed the vital role that social media plays in communicating information about real-time conditions with examples drawn from the Los Angeles Carmageddon project and Hurricane Sandy. While these tools have enormous potential to increase the availability of real-time information, they also come with potential risks and must be managed carefully to ensure accuracy.

REFERENCES

American Public Transportation Association. 2012. *Transit Ridership Report, Second Quarter 2012.* http://www.apta.com/resources/statistics/Documents/Ridership/2012-q2-ridership-APTA.pdf

Austin Mobility website. http://austin-mobility.com/

Brenner, J. 2012. Pew Internet: Social Networking. Pew Internet & American Life Project. http://pewinternet.org/Commentary/2012/March/Pew-Internet-Social-Networking-full-detail.aspx

Brown, R. 2011. Rebelations. http://www.rebelbrown.com/

BWI Marshall Facebook page. http://www.facebook.com/pages/BWI-Thurgood-Marshall-Airport-BWI/101280343252797

CADMV. 2010. Five Ways to Save Time at the DMV. *DMV Blog*, November 12. http://cadmv.wordpress.com/2010/11/12/five-ways-to-save-time-at-the-dmv/

CADMV. 2011. DMV Social Media Pilot Program Results. *DMV Blog*, May 5. http://cadmv.wordpress.com/2011/05/05/dmv-social-media-pilot-program-results/

Cecconi, E. 2012. Case Study: How To Create Inspiring Social Media Content for an Airport. htttp://simpliFlying.com/2012/inspiring-airport-social-media-content/

Evans-Cowley, J. S. and J. Hollander. 2010. The new generation of public participation: Internet-based participation tools. *Planning Research and Practice*, 25, 397–408.

Evans-Cowley, J. and G. Griffin. 2012. Micro-participation: community engagement in transportation planning with social media. *Transportation Research Record*.

Falls, J. 2010. What is engagement and how do we measure it? http://www.social-mediaexplorer.com/social-media-marketing/what-is-engagement-and-how-to-we-measure-it/

Frei, C. and H. Mahmassani. 2011. Private time on transit: dimensions of information and telecommunications use on Chicago transit riders. http://www.transportchicago.org/uploads/5/7/2/0/5720074/frei_tc_website_posting.pdf

Golden, M. 2011. *Social Media Strategies for Professionals and Their Firms: The Guide to Establishing Credibility and Accelerating Relationships*. Hoboken, NJ: John Wiley & Sons.

Goodspeed, R. 2010. The dilemma of online participation: comprehensive planning in Austin, Texas. http://web.mit.edu/rgoodspe/www/papers/RGoodspeed-Austin_Online_Participation_9-19-10.pdf

Hendel J. 2011. Farragut Crossing wins Metro's virtual tunnel name game. http://www.tbd.com/blogs/tbd-on-foot/2011/08/-farragut-crossing-wins-metros-s-virtual-tunnel-name-game— 12429.html

Jacksonville Transportation Authority. 2011. Name New Fare Card Contest. http://www.jtafla.com/News/showPage.aspx?news=186

Lindsay, B.R. 2011. Social Media and Disasters: Current Uses, Future Options, and Policy Considerations. Congressional Research Service Report for Congress.

Los Angeles County Metropolitan Transportation Authority. 2012. Westside Subway Extension: Final Environmental Impact Statement/Environmental Impact Report. http://www.metro.net/projects/westside/final-eis-eir/

Metropolitan Transportation Authority via Flickr. http://www.flickr.com/photos/mtaphotos/8138916414/in/set-72157631889343888

New Media Lab. 2009. Top 10 Cities Using Twitter. http://nmlab.com/social-media/top-10-cities-using-twitter/

Noor Al-Deen, H.S. and J.A. Hendricks. 2012. *Social Media: Usage and Impact*. Lanham, MD: Lexington Books.

Passenger Focus. 2011. Information: Rail Passengers' Needs during Unplanned Disruption. http://www.passengerfocus.org.uk/research/publications/information-rail-passengers-needs-during-unplanned-disruption

Perk, R. 2012. New poll finds that people hate traffic, love transit. *Switchboard: National Resources Defense Council Staff Blog*. http://switchboard.nrdc.org/blogs/rperks/new_poll_finds_that_people_hat.html#r

Reuters. 2011. CDC 'Zombie Apocalypse' disaster campaign crashes website. http://www.reuters.com/article/2011/05/19/us-zombies-idUSTRE74I7H420110519

Schor, E. 2009. LaHood's Twelve-Word Definition of 'Livability'. *DC Streetsblog*. http://dc.streetsblog.org/2009/10/05/lahood-defines-livability-in/

SEPTA Twitter account. @SEPTA. https://twitter.com/SEPTA/status/306810576733085697; https://twitter.com/SEPTA/status/306753568940118016

SNAPP website. http://snappatx.org

SNAPP blog. http://blog.snappatx.org/

SNAPPatx. 2010. Comment on Young People on Car Ownership. http://blog.snappatx.org/2010/11/young-people-on-car-ownership.html#comment-form

Social Media Performance Group. 2011. How to Engage with Social Computing. *Social Media Performance Group's Blog.* http://smperformance.wordpress.com/2011/10/21/how-to-engage-with-social-computing/

Stuef, J. 2012. The Man behind @ComfortablySmug, Hurricane Sandy's Worst Twitter Villain. *BuzzFeed FWD.* http://www.buzzfeed.com/jackstuef/the-man-behind-comfortablysmug-hurricane-sandys

Sweet, S. and M. Pitt-Catsouphes. 2010. *Talent Pressures and the Aging Workforce: Responsive Action Steps for the Transportation and Warehousing Sector.* Boston: Sloan Center on Aging & Work at Boston College.

Taylor, B.D., E.J. Kim, and J.E. Gahbauer. 2009. The thin red line: a case study of political influence on transportation planning practice. *Journal of Planning Education and Research,* 29, 173–193.

Tinker, T. and D. Fouse. 2009. Expert Round Table on Social Media and Risk Communication During Times of Crisis: Strategic Challenges and Opportunities. *Booz Allen Hamilton.* http://www.boozallen.com/media/file/Risk_Communications_Times_of_Crisis.pdf

Tinker, T. and E. Vaughan. 2012. Risk and Crisis Communications: Best Practices for Government Agencies and Non-Profit Organizations. *Booz Allen Hamilton.* http://www.boozallen.com/media/file/Risk-and-Crisis-Communications-Guide.PDF

John A. Volpe National Transportation Systems Center. 2010. Current Uses of Web 2.0: Applications in Transportation. Case Studies of Select Departments of Transportation. Washington: John A. Volpe National Transportation Systems Center.

Washington State Department of Transportation. 2012. SR 99 tunneling machine tweets her name: Bertha. http://www.wsdot.wa.gov/News/2012/12/10_sr99_tunnel-machine-name.htm

161

5

Learning from Customers and Community

Besides facilitating better connections between transportation organizations and their constituents, social media can help agencies work with users to make a system better, evaluate system use, and analyze customer opinions and perspectives. Social media analysis tools are pushing transportation agencies into a new age in which frequent and occasional customers can have a greater say and more positive impact on the transportation system.

In this chapter, researchers from four universities share details about their work on the forefront of transportation-related social media research. The first section was written by Aaron Steinfeld, John Zimmerman, and Anthony Tomasic, researchers at Carnegie Mellon University who created the Tiramisu mobile application. Here they explain the concept of social computing, often called crowdsourcing or human computation. In the next section, Stacey Bricka, Debbie Spillane, and Tina Geiselbrecht, all from the Texas Transportation Institute, and Tom Wall of Georgia Tech explain how transportation surveys are evolving to include social media as a means to reach respondents and understand social networks. Finally, Satish Ukkusuri, Samiul Hasan, and Xianyuan Zhan of Purdue University discuss data analysis techniques as social media move into the future with methods such as data mining for user perception and activity pattern recognition.

BRINGING CUSTOMERS BACK INTO TRANSPORTATION: CITIZEN-DRIVEN TRANSIT SERVICE INNOVATION VIA SOCIAL COMPUTING

Aaron Steinfeld, John Zimmerman, and Anthony Tomasic

The development and rapid expansion of high-speed residential connectivity, online social networking services, and most recently smartphones have changed the world and given birth to *social computing*, a new kind of computing in which large groups of people and computer systems do things in collaboration that neither can do alone. Interestingly, this rapid change in communication technology has led government agencies to reconsider how they might communicate and engage with the public.

Examples of exciting new services include transit agencies that provide up-to-the-minute information via Twitter, Facebook, blogs, and consumer-friendly mobile GIS tools. This section focuses on this new technology and includes some definitions, a review of trends, and a case study showing what citizens can do in collaboration with their transit agency.

Social Computing and Needs for Public Services

Urban planning has a long history of considering the needs of citizens when designing and altering local environments (Brabham 2009). Planners most often use surveys, focus groups, and community meetings to achieve some measure of citizen participation. However, planners note that these traditional methods can be slow, expensive, and often engage only a tiny fraction of citizens.

Recently, planners have begun to investigate the use of social computing to improve the efficiency and effectiveness of citizen participation (Brabham 2009). Community participation approaches to the use of social computing can be divided into two categories: general-purpose and domain-specific. The use of general-purpose social media for community engagement is explored earlier in this book, and later sections of this chapter discuss the evolving uses of social media in surveys and data mining in greater depth.

General-purpose social media services such as Twitter and Facebook can tap large numbers of participants due to their huge popularity. However, they often lack features that meet specific data needs, such as the ability to connect information to specific places and pieces of infrastructure. Domain-specific approaches can more easily collect specific

164

information that better maps to the issues at hand. However, these smaller scale systems make building a critical mass of participants much harder.

Social computing provides a new technology platform that can bring large groups of people together around a topic or issue. The ability to mix large groups of people and the power of computational devices and systems has created services never before imagined, such as Wikipedia, YouTube, and Twitter. These services work because of the ongoing collaboration of people and technology. In the research community, the concept of uniting large groups of people and computing has been expressed in several different forms, including:

- *Citizen science* allows people to work as sensors and relay information to scientists or advocacy groups, increasing the reach of science (Silvertown 2009). Examples include the Audubon Christmas day bird count that has been underway since 1900. A more recent example can be found in the Safecast service (http://safecast.org). After the Fukushima Daiichi nuclear plant accident in Japan, Safecast collected Geiger counter readings shared by citizens and displayed visualizations of the information on an online map.
- *Crowdsourcing* combines the concepts of "outsourcing" and the "wisdom of crowds" to show that many people working together are smarter than a few people working in isolation (Howe 2006). Examples include the content recommenders found at Amazon and Netflix; Wikipedia, a community-constructed encyclopedia; Twitter, where the comments of users reveal unfolding trends; and Mechanical Turk, which makes thousands of workers around the world available to perform small tasks.
- *Human computation* views people as an informational resource that computational systems can use to get smarter (Von Ahn et al. 2008). These systems are often designed as games in which people's play provides valuable information that machine learning systems use to extend their understanding of the world.
- *Participatory sensing* views the ever-increasing number of mobile phones in the world as a new type of instrumentation and a new source of sensor information (Burke et al. 2006). Modern smartphones can sense their locations, movements, and noise and light levels. People with smartphones can easily share their observations of unfolding situations from civil unrest to blizzards.

In the abstract, these overlapping social computing forms have little value to transportation service providers. However, considering the inputs and outputs of the different forms can open up possibilities for new services. For example, human computation is typically most successful when structured as a game in which the players do not need connections to the informational goal of the system. Similarly, participatory sensing systems are typically embodied either as games (e.g., Foursquare) and/or they run as background processes invisible to the users of mobile devices.

The Inrix service automatically receives inputs from thousands of GPS devices installed on local and long-haul trucks and on smaller fleet vehicles. Bringing these many location traces together allows the service to estimate traffic flows across the U.S. This participatory sensing approach leverages a dataset and technical infrastructure that is already in place to provide a valuable new service.

The Waze* human computation service offers traffic information collected from a large group of people, but it uses a game metaphor to drive participation. When data is needed for a specific area, the service places a game icon on the specific road link and awards points to Waze users who drive the segment. Although both approaches can be useful, not all problems can be solved by invisible background actions or by tempting people to engage in game play.

Dozens of transit agencies use Twitter, a general-purpose social computing tool, to provide domain-specific communication. Many use Twitter to push out just-in-time information on events such as unexpected closures and detours. Riders who access these live updates can adjust their travel plans accordingly. The Port Authority of Allegheny County (PAAC) provides an interesting example. Two major disruptive events in Pittsburgh wreaked havoc on the transit service: the 2009 G20 meeting with its assorted protests and security actions and a storm that dumped three feet of snow in February 2010. PAAC used Twitter heavily during these events, often copying and pasting internal messages straight into its tweet stream (Schwartzel 2010).

Interestingly, a two-way flow of information emerged when PAAC's Twitter followers sent tweets to the agency with reports of what they saw in their neighborhoods. These disruptions and the access to live information via Twitter led to large increases in PAAC's followers on the platform (Steinfeld et al. 2012a).

The use of social media for real-time communications was discussed in the previous chapter. Going beyond real-time communication, some researchers suggest using Twitter as a source for gathering real-time data

---

* http://www.waze.com

from populations during evacuations and public planning (Brabham 2009; Turner et al. 2010). Recently, social interaction within mobile transit information applications has become more common. Several transit applications such as Roadify[*] and naonedbus[†] now support rider–rider dialogues.

Some researchers, including the authors of this section, have explored how rider-to-rider communication can drive transit service innovation (Yoo et al. 2013). Work by Machado et al. (2012) suggested that entertainment and content sharing during a commute can reduce perceptions of a trip's duration. The authors' work suggests that the combination of rider–rider dialogue along with location- and route-specific alerts from an agency can significantly increase the appeal of using transit (Yoo et al. 2010; Steinfeld, et al. 2012a, Yoo et al. 2013). Incorporating communication into transit information services has the added benefit of improving data analytics, as discussed later in this chapter, due to the ability to match rider behavior to social media content.

eGovernment and Social Computing

For several years, government agencies have been investigating how to better use new communication technology. Many eGovernment initiatives have focused on making services or information about government decision making available on the web. More recently government agencies and third-party developers have started looking toward social computing as a way to enhance these activities and to increase citizen engagement.

An example of this approach is the White House petitions site.[‡] It is worth noting that this can lead to unexpected citizen actions such as the recent petition to build a "Death Star" by 2016.[§] Not surprisingly, these systems encounter many of the same challenges faced by more traditional public involvement techniques.

Other third party tools have been created to increase government–citizen communication in planning via social computing. OpenPlans' Shareabouts[¶] provides a good example. This service enables participatory urban planning by allowing members of a community to mark locations

---

[*] http://www.roadify.com
[†] https://play.google.com/store/apps/details?id = net.naonedbus
[‡] https://petitions.whitehouse.gov
[§] https://petitions.whitehouse.gov/petition/secure-resources-and-funding-and-begin-construction-death-star-2016/wlfKzFkN
[¶] http://openplans.org/work/shareabouts

where they would like to see changes on an online map. The service is typically embedded within a specific planning campaign website and has mostly been utilized to collect feedback to inform the design of bike sharing services (New York, Portland, Cincinnati, etc.). The service has also been used for other purposes such as documenting dangerous intersections (Brooklyn) and problematic phone booths (New York).

The Urban Mediator offers another interesting example of citizen participation in digital planning (Saad-Sulonen et al. 2012). This online tool allows citizens and urban planners to create and share topics. It has been used in Finland to engage citizens in discussing traffic issues and share their ideas about design proposals. Like Shareabouts, Urban Mediator provides evidence that citizens can and will engage in planning discourse related to local services.

There have been several third-party services designed to connect citizens and government services. For example, SeeClickFix* and FixMyStreet.com (King and Brown 2007) both allow people to report problems like potholes to municipal authorities. SeeClickFix goes a step further by providing tools that allow municipal agencies to feed reports directly into their reporting systems for routing to the responsible department. ParkScan[†] is similar in this regard. This service allows city residents to report on problems with their parks and receive feedback from the local maintenance staff about the steps taken to address issues. In addition to this feedback loop, the service provides rich evidence of citizen involvement that park services used to lobby for new resources. The authors' research suggests the feedback loop is critical; without an agency response, users may believe that their concerns are simply disappearing into a black hole (Yoo et al. 2010).

Service Design

The transition of most western countries from manufacturing to service economies spurred the growth of a new discipline called service design. Service designers work to develop a *service concept:* a description of a system that meets the needs of customers and fulfills a service provider's strategic initiative (Goldstein et al. 2002). Interestingly, almost all technology-driven innovation follows a product-centric design process of making *things* to sell to customers. However, this is beginning to change in the software development community.

---

* http://seeclickfix.com
† http://www.parkscan.org/

168

A good example of this transition can be seen in software startups that offer web-based services. Traditionally, each small startup had to purchase and maintain its own server. However, today most new web startups contract with service providers to host their systems, paying more or less based on the volume of traffic and resources that they use. This has significantly reduced the initial capital investment needed to get a startup running and it allows new businesses to rapidly scale their growth based on actual demand.

With products, value is exchanged at the time of purchase; ownership is transferred from the maker to the consumer. With services, value emerges during a customer's interaction with a service provider. There is no exchange of ownership; instead value is co-produced between the service provider and the customer. Service design utilizes the idea that customers have competences, and these competences can be sources of value for service providers (Prahalad and Ramaswamy 2000). As an example, customers commonly share their competence in the software industry by becoming beta-testers. These customers get a sneak preview of what is coming and some influence over the design of the new version. The software companies in return get free labor to test their software.

From this perspective, service design requires design teams to investigate what customers can do for the benefit of the service provider. Applying the service design process to drive innovation at transportation agencies, the agencies should be asking not just what they can do for citizens, but what citizens can do for themselves and for the good of the transportation agency. This type of investigation almost always results in a host of new programs or services.

The service design community recognizes public services as a special case. Public services often function as monopolies and are therefore not driven to innovate through competition. This fact is not necessarily a negative quality, as competition can diminish the importance of marginalized communities like the poor and people with disabilities, who are often the intended audiences for public services. In fact, many public services exist specifically to address issues of fairness by creating a more level playing field for marginalized communities (Boyne 2003). However, the lack of direct competition can contribute to the slow adoption of new systems and tools. One example is the reluctance of many transportation agencies to leverage free social computing communication services such as Facebook and Twitter.

Many public services have looked to information technology as a way to improve service delivery. Traditional technology approaches focused on automation and improving the speed of service delivery. However, agencies are beginning to learn that their customers' main desires are often for different services, not for automated services (Bradwell and Marr 2008). When detailing opportunities for public service improvement, Boyne (2003) noted seven metrics for assessing service improvement and innovation:

- **Quantity of outputs** — More services offered
- **Quality of outputs** — Speed and reliability
- **Efficiency** — More outputs for less money
- **Equity** — General sense of fairness toward citizens in terms of costs and benefits
- **Outcomes** — Increased usage of a service by citizens or by a percentage of citizens
- **Value** — Cost per unit of outcome
- **Consumer satisfaction** — Often a result of several of metrics above

Service researchers have proposed that engaging end users to design services with a provider (co-design) may be the best approach to drive public service innovation. They noted three positive outcomes (Bradwell and Marr 2008):

1. Services that are more responsive to the changing needs of citizens
2. Increased trust of the government through citizens' more positive engagement with services
3. Increased social capital through an increased sense of community

The challenge now is how to best engage citizens in collaborating with transportation agencies to design new services that they want and are willing to pay for through taxes, tolls, fares, and other user fees.

Case Study: Tiramisu

The research team at Carnegie Mellon University (CMU) investigated how social computing might help citizens to co-produce value with their transit services. Researchers developed a system called Tiramisu—which means "pick me up" in Italian—that allows transit riders to create a real-time arrival information system through crowdsourcing. In other words, users share location traces and reports of vehicle fullness from their smartphones as they take transit trips (Zimmerman et al. 2011; Steinfeld et al. 2012a).

The application is especially beneficial in transit systems where real-time vehicle location information is not available to passengers. Instead, in collaboration with the transit provider, the riders collectively generate information about vehicle arrivals. Collecting this information from riders lessens the need to install expensive commercial automated vehicle location systems. Tiramisu also allows riders to share kudos and complaints, providing the transit system with a new source of information on how customers experience its service.

The research for this project began with a focus on how transit riders might function as sensors across the transit service by reporting delays, crowding, and other breakdowns in infrastructure as they occurred. However, interviews with transit riders revealed that they were much more interested in knowing if they had missed their buses than in reporting system problems (Yoo et al. 2010). Unfortunately, real-time arrival data based on automatic vehicle location (AVL) is expensive to collect and disseminate. One analysis estimates the cost for a basic system serving a mid-sized city with an 800-vehicle fleet at $16.5 million (Parker 2008). When asked, transit agencies often cite the high capital and operating expenses associated with deploying proprietary real-time arrival data systems as the primary reason for not providing this information.

Taking a service design approach, members of the team began to view the smartphones carried in the purses, pockets, and backpacks of many transit riders as sources of customer competence. These devices are equipped with location awareness, a network connection that has already been paid for, and screens that support easy input and output of information. Designers began to speculate on how they might convince transit riders to create their own AVL system using technology they already carry.

Another primary concern of the project was accessibility, particularly how social computing might be leveraged to improve service experiences for blind, mobility impaired, and elderly riders who often have a greater dependency on public transit. In addition to service design, the team followed a universal design process. This type of design focuses on making systems and services accessible to all users, not developing special systems intended to serve only people with disabilities. From a universal design perspective, a real-time information system generated from the efforts of all riders could significantly improve the experiences of disabled riders; it takes considerably more effort for these passengers to travel to a transit stop and they can be more at risk if they miss their vehicles.

Tiramisu allows users to find nearby bus stops. They start by pressing the "Nearby" button on their home screens. Figure 5.1 shows the

Figure 5.1 A series of screenshots from the Tiramisu application shows various options for users. (From: Carnegie Mellon University. With permission.)

progression of screens in the application. The button transitions them to a map where they can select the stop they wish to depart from and a list of arrival times for this stop. Here the system provides predictions. If another Tiramisu user currently shares a location trace, the arrival time is noted as a "rider real-time prediction." If no one is currently sharing but several riders have shared traces for this trip in the recent past, the arrival is noted as a "rider historical estimate." When neither real-time nor historical predictions are available, the system displays an arrival time based on the service's schedule. Both the historical and the real-time estimates also include predictions of how full vehicles are expected to be.

As the vehicle arrives, a user selects the route to be boarded from a list. He or she then chooses a destination from a list of all possible destinations for the selected route. The system asks for a destination in order to know when to stop sharing the location trace. Finally, the user indicates how full the bus is and presses the "Start Recording" button to share a location trace from his or her phone. Details on the design process, system architecture, and rationales for various features were carefully considered throughout the process (Yoo et al. 2010; Steinfeld et al. 2012a; Zimmerman et al. 2011).

An early version of the system was used in a field trial with riders on the Port Authority of Allegheny County in Pittsburgh (Zimmerman et al. 2011). The team wanted to assess whether users would be willing to share location traces while riding, discover software bugs, improve prediction times, and understand how a mobile application affects the transit experience. Riders with disabilities were not explicitly recruited and none participated. Since accessibility was a critical goal, even this preliminary version fully supported screen reading using the iPhone's VoiceOver feature. The publicly deployed iPhone version of Tiramisu is also VoiceOver-compatible.

The field trial included data from 28 people over a span of 38 days. Participants returned 2 weeks after they started to use the system to complete a survey and receive token payments. Participants were recruited from a specific region of Pittsburgh to increase the chance that their contributed data would directly benefit other participants. This appeared to be the case; sampling of stops within this transit corridor frequently revealed at least one real-time and several historical predictions. Researchers conducted an analysis looking at the first 21 days of use for each participant (they started on different days). Participants shared location traces for approximately 56% of the trips they took. Approximately 13% of the arrival estimates they viewed contained real-time or historical predictions (2,132 of 16,263). This was a very good rate based on the number of possible buses and the limited number of participants.

The goal of having users share comments about the service with the agency was fulfilled. Most participants indicated that they had never contacted a transit agency with a compliment or complaint prior to their participation in this study. Interestingly, 14 participants submitted a total of 22 reports to the transit system, thereby showing increased rider investment in overall service quality.

The Tiramisu system has been in public operation in Pittsburgh since late July 2011. As of this writing, users have shared more than 68,000 location traces. The service was recently released for Syracuse and Brooklyn and more cities are planned. A preliminary analysis of the public deployment in Pittsburgh is currently under review. Analysis of the usage log files suggests that crowdsourcing arrival information is working.

It is worth noting that the vehicle fullness and rider origin destination data collected by Tiramisu can be extremely difficult to collect using traditional transportation information systems. For example, most transit agencies only know how many people enter and exit a bus; they most often have no way to link the origins and destinations of individual riders. Transit agencies traditionally estimated crowding by assigning individuals to a location along a route to estimate the level of fullness as buses drove past. Empowering customers to collect this data saves money for agencies in terms of staff resources and also has the potential to provide more complete and accurate data for use in route planning and daily operations.

The first 1,000 notes shared in Pittsburgh were analyzed to determine trends and underlying themes, and the results of this analysis are reported in previous papers (Steinfeld et al. 2012b). As with the field trial (Zimmerman et al. 2011), users of the deployed system provided notes to the transit agency at rates much higher than would normally be expected. Many of the notes pointed to negative service experiences and cited missing or late buses. Other comments noted positive and negative feedback about drivers, maintenance issues, and thoughts about fellow riders.

These notes suggest a willingness among users to report on unfolding conditions and situations during their trips and at their transit stops. Interestingly, the system captured instances in which drivers did not allow riders using wheelchairs to board the bus. This confirms laboratory results that implied riders would use mobile reporting to document such events (Steinfeld et al. 2010).

Conclusions

As is clear from the various examples provided here, social computing techniques provide a wide range of interesting and novel ways to engage riders. Transportation agencies should consider the following when considering and pursuing a social computing approach:

- At a minimum, use free services like Facebook and Twitter to push out real-time updates and other time-sensitive information.
- Offer an interactive dialog system allowing citizens to share details of their transportation experiences as they happen, whether they are waiting for a bus or stuck in traffic. Have a

174

feedback mechanism so participants know that the agency heard their issues and concerns and maybe addressed them.

- Offer a user-to-user communication channel. This can be especially helpful for transit agencies, where riders can brainstorm to create new services such as reporting on bicycle parking availability at stations. Monitor this channel to see what stakeholders want to talk about.
- Leverage customers and transportation workers as powerful sensors. Give them tools that allow them to collect information and report back to the agency, such as the mood on a vehicle or a waiting time at the department of motor vehicles.
- Use social computing to create a dialog with citizens and critical stakeholders (e.g., local businesses) about the state of the current system and ideas for new designs. Collect "likes" and comments on social media channels as evidence of citizen desire for new services the agency may want to consider when working on future funding measures.

Clear challenges surround any new customer interaction approach, but the opportunities to create new information channels, gather data, and tap into rider expertise are great.

MAY WE HAVE A FEW MINUTES OF YOUR TIME? USING SOCIAL MEDIA IN TRANSPORTATION SURVEYS

Stacey Bricka, Debbie Spillane, Tina Geiselbrecht, and Thomas Wall

Surveys are common tools for transportation planning and research. They vary from short polls to assess attitudes and opinions to more detailed surveys about traveler origins and destinations. They may collect data on demographics; transportation modes; trip characteristics like location, time of day; and trip purpose. Survey findings can help transportation agencies plan new projects from a new airport ground access service to a transit line extension.

The survey research industry is working to incorporate the benefits of mobile technology into survey techniques, and online and smartphone-based surveys are becoming commonplace. Using social media to conduct surveys has proven more of a challenge, however, and to date social media is used most frequently to communicate details about a survey to the general public or enlist participation in online panels and informal polls.

175

These uses of social media in such surveying efforts and more advanced techniques for incorporating social media into surveying are highlighted in this section.

This section is organized into three parts: (1) how the general survey research industry uses social media, (2) how social media is used currently in transportation surveys, and (3) potential uses for social media in transportation surveys.

Social Media in Survey Research

The methods used to conduct transportation surveys have evolved over time but closely follow trends in the survey research industry. Origin–destination surveys, in particular, pay special attention to specific travel patterns and are frequently used to plan new services or to update regional travel models. Accordingly, these surveys require scientifically selected random samples that can be adjusted statistically to reflect the demographic characteristics of the study area. These surveys differ from those that are advertised for the general public to complete without any type of sampling and are commonly called non-probability or "choice" samples. The requirement for random sampling for many transportation surveys has resulted in limited application of social media in transportation surveys.

The survey research industry as a whole is interested in leveraging technology to improve survey participation. In addition to smartphone-based and online surveys, advances in technology have enabled the survey community to develop more advanced and targeted sampling frames that are critical components in developing a statistically valid survey.

Social media, on the other hand, is broadly defined as "a phenomenon that stretches from blogs to wikis, from communities to YouTube" (Poynter 2010). Although social media platforms have a broad reach, which would seem ideal for drawing a survey sample, they have several drawbacks that currently limit their usefulness in scientific survey efforts.

As of this writing, it is known that the universe of social media users does not reflect the demographics of the U.S. In 2012, U.S. social networkers skewed more female and younger than the general U.S. population. Some 54% of social networkers were female, compared to 51% of overall population aged 12 and older, and while 21% of the U.S. population was aged 12 to 24, almost 1 of 3 social networkers (32%) fell into that age range (Edison Research 2012). Because personal details about social media users are often confidential and may not even be true, it is currently impossible to draw a social media sample using standard demographic characteristics like gender, age, and geographic location.

Because of the need to resolve the qualitative nature of social media with the quantitative underpinnings of survey research, the use of social media to conduct a survey has been limited. Within a few years, however, it is anticipated that researchers will identify ways to use social media to conduct surveys. In the meantime, they see great opportunities to leverage the power of social media to promote survey opportunities (Bassett et al. 2012; Meagor n.d.). Links to web surveys are routinely promoted via Facebook posts, Twitter tweets, and other social media platforms. In addition, social media can be used to test various aspects of a survey. According to Meagor, social media can enhance brand awareness and reputation, engage the public, and provide useful feedback. Meagor also suggests that social media be used to test survey questions, survey branding, and other aspects of a survey effort.

Henning (2010) recommends incorporating social media into survey design in the following areas:

- **Study design** – Reading social media posts can help researchers better understand the target market and frame the research hypothesis.
- **Questionnaire design** – By reviewing social media conversations, researchers can observe how people talk about the topic at hand and use this information to frame questions in the language of potential respondents. For example, researchers are interested in "origins and destinations of travel" but the general public just thinks about the places they go on a given day. Researching social media posts can also help researchers develop relevant and comprehensive choice lists for closed-end questions.
- **Fielding the survey** – For surveys in which a choice or non-random sample is appropriate, researchers can publish links to an online survey through Twitter, Facebook, and other social media platforms.
- **Survey analysis** – Researchers can supplement quantitative surveys with online social media commentary.

Social media has also proven useful in enlisting participants for online survey panels (Meagor n.d.). These panels are useful when a survey sponsor is interested in polling the same group of respondents on a regular basis for qualitative data. Targeted advertisements (Smart-survey blog, 2012; Swicegood 2012) can be used to attract certain types of social media users to receive these regular surveys. While this results in the creation of a choice sample (Yu and Kak n.d.), it can be employed alongside a statistically random sample to provide detailed insights about specific population groups of interest.

The use of social media to develop a sample, however, should be pursued with caution. According to DeVall (n.d.), the quality of such online panels drawn solely from social media may result in a biased sample. This issue may be resolved in part by using a vetting process to screen out professional survey participants or those with alternative agendas.

In some instances, however, the vetting process may be facilitated by the use of social media. For example, blogs and wikis often have membership requirements, and online social networking sites such as Facebook and LinkedIn enable users to associate with specific networks (e.g., geographical, place of employment). These mechanisms of the social media sources can be used to target surveys toward specific groups or screen out groups that may introduce bias into a sample.

In sum, in the general field of survey research, social media is not used currently to conduct scientific surveys. It is used to support surveys by testing branding and marketing messages as well as testing survey questions and other aspects of survey design. In addition, social media is useful in promoting qualitative surveys—those that are not generalizable to the public or anticipated to be useful in a scientific study. Finally, social media presents the potential to allow the creation of non-scientific respondent groups representing demographic or other characteristics of interest.

Social Media in Transportation Surveys

Transportation researchers are actively seeking opportunities to use social media in surveys, but they are proceeding cautiously. A recent analysis of the use of social media by travel survey participants was conducted as part of a regional travel survey for the New York–New Jersey metropolitan area. As part of the pilot test for that survey, the team noted that "nearly all of the interviewees were Facebook users and familiar with other social media sites." Respondent perspective on the use of social media for these types of surveys suggests that the potential exists for creating "awareness of and interest in a ... survey effort" (Chiao et al. 2011).

Indeed, the main use of social media in this context is to promote transportation surveys. However, the more researchers understand and can identify how to leverage social media, the more the media will be used to support survey research. Examples of current applications and possibilities for future transportation survey uses are explored in this section.

Scientific Transportation Survey Promotion

Because most transportation surveys are publicly funded, they are usually accompanied by some type of press activity. A reporter may attend a board meeting where the survey is discussed and subsequently write about the effort for a newspaper, summarize the details in a blog post, or share the information via Twitter. Frequently the transportation agency sponsoring the survey issues the details directly as a means of legitimizing the survey and encouraging participation via a combination of traditional and new media.

Promoting a survey via social media uses the same format as a traditional press release: reason for survey, who will be asked to participate, and name of a contact who will provide additional information. For example, in the spring of 2011, the Southeastern Pennsylvania Transportation Authority (SEPTA) in Philadelphia announced that the agency would be conducting an on-board survey. SEPTA used its Facebook page to tell riders which lines would be surveyed, summarize the reasons for the survey, and ask them to participate if they received a survey. SEPTA's (2011) post read, in part:

> The survey's goal is to study reasons for ridership, travel patterns and analyze customers who transfer between two or more services. If you receive a survey, take a moment to answer the questions and return it. It helps for future service planning in the five counties we serve and the overall region.

Reader comments reflected a mix of support for the survey and general frustration with the agency. Sample posts included the following:

Respondent 1: Happy to have been able to fill out the survey on the Trenton line this morning! Thanks

Respondent 2: I took the survey, and couldn't find enough room on the thing to write all of my feedback. I wish the survey was more oriented toward customer satisfaction. It seemed primarily filled with pragmatic questions. I had to write my concerns in the margins.

SEPTA took the opportunity to respond to the latter complaint and wrote, in part, "You can always use www.septa.org/comment to send us detailed comments. We also look for feedback, on a variety of topics, using our website survey feature at www.septa.org/cs/survey/. Send in your ideas if you have a topic you want us to consider for a future survey."

Some transportation agencies used online videos to encourage participation in surveys, especially household travel surveys. Throughout the U.S., travel surveys are conducted to create snapshots of regional travel. A random sample of regional households is contacted and asked to document all trips in a 24-hour period made by all household members regardless of mode. Completing these surveys can take a household at least an hour on average and more for a larger household. To encourage participation, some agencies created and posted videos to YouTube. The videos focus on the value of the household's information to the regional transportation planning process, and their use is becoming common in large-scale regional travel surveys.

In Minneapolis–St. Paul, regional planners at the Metropolitan Council created a video to introduce citizens to the 2010 Travel Behavior Inventory (TBI),* a series of surveys conducted every 10 years to provide policymakers and researchers with data about travel patterns in the Twin Cities region. Posted on YouTube, the video described the goals for the TBI, summarized the survey process, and reassured participants that no personal information would be shared.

Similarly, in 2012, the California Department of Transportation posted a YouTube video to inform citizens about the California Household Travel Survey.† The video used a former local television reporter to demonstrate the process of using a travel diary and/or a GPS unit to document daily behavior—from driving to the train in the morning to stopping at the grocery store on the way home from work. Interviews with local transportation officials focused on the importance of the data in developing area transportation programs and policies.

Qualitative Survey Promotion
Agencies often develop short surveys that are open to the general public to engage them in a transportation-related discussion. These are considered qualitative surveys because they are not supported by scientific sampling. Instead, people can opt to participate if they are interested in the topic. The details obtained by these surveys are often used in conjunction with qualitative data obtained from public meetings on upcoming transportation projects. Despite their non-scientific nature, these surveys are useful in explaining the more quantitative modeling results or fine-tuning

---

* http://www.youtube.com/watch?feature=player_embedded&v=bALKrAjaX3Q0
† Caltrans Survey. http://www.youtube.com/watch?v = h1KjCZQaDJ8

Figure 5.2 The Potomac and Rappahannock Transportation Commission used a splashy graphic on its Facebook page to encourage teenagers to participate in an online survey. (Screen capture from www.facebook.com/prtctransit.)

public outreach messages. Transportation agencies almost always use social media to promote these survey links.

In the fall of 2010, the All Campus Commuter Board (ACCB) at the University of California, San Diego posted the link to a survey on a Facebook page. The text announcing the survey first referenced recent changes in parking and shuttle services, then asked for participation in the survey to reflect commuter opinions about transportation options. A link to the actual survey was provided.*

Similarly, the Potomac and Rappahannock Transportation Commission advertised a Facebook survey for riders who purchased and used Teen Summer Passes in 2012. The purpose of the survey was to gain input from users to improve the program. The commission offered an incentive of a $100 Visa gift card to boost responses (Figure 5.2).

In June 2011, TriMet (the transit provider in Portland, Oregon) conducted a survey to guide its use of social media (Rose 2011). The agency

---

* All Campus Commuter Board (ACCB). http://www.facebook.com/events/1230411277 50295/

Figure 5.3 The North County Transit District used Twitter to help promote a regional transportation survey. (Screen capture from @gonctd.)

used Twitter to invite participants with the following tweet: "Hey TriMet Twitter followers, tell us what you think about how we use Twitter. Take our survey."

Transportation organizations often take advantage of the viral nature of social media to promote surveys by encouraging their agency partners to post or retweet links to general public surveys. For example, when the San Diego Association of Governments updated its regional coordinated plan, it conducted an online survey of the transportation needs of residents in the rural areas of San Diego County. The North County Transit District that serves part of San Diego County used Twitter to encourage its followers to participate (Figure 5.3).

Similarly, Capital Metro, the Austin area transit provider in Texas, used Facebook to promote a general public survey conducted by the Texas A&M Transportation Institute (TTI). Such inter-agency marketing helps expand the reach of a survey and increases the legitimacy of the research in the eyes of potential respondents.

Developing Online Panels
As noted above, some transportation agencies use social media to assemble a panel of citizens to be polled on a variety of topics over a set period that

can last from a few months to a year. For example, when the Minnesota Department of Transportation established an online community to provide ongoing customer feedback and interaction, it used social media to help recruit a panel of 600 residents (Wallace 2012). Similarly, Houston Metro used social media to announce that it was taking applications for a citizen's advisory committee (Sit 2012). A post on the Metro blog asked riders and non-riders alike to consider applying.

When assembling panels, researchers implement controls to ensure that the groups are representative of the target population. Once the panels are established, agencies frequently poll the respondents and ask them to participate in short online surveys. Communications with online panels can take place via e-mail, texting, or a specific website; in addition, special Facebook or Twitter groups can be established for communication with panel members.

Potential Uses for Social Media in Transportation Surveys

While current applications are still limited, social media presents great potential for transportation surveys and customer-based research. Possible uses include testing survey design, identifying respondent groups, and using social media data to inform transportation modeling.

Testing Survey Design

One area that has not yet been tested, but seems well suited to a social media application, is testing survey branding, survey questions, and other aspects of survey design prior to the launch of a scientific or qualitative survey effort. Whether through the establishment of an online panel or just asking questions of social media followers for a transportation agency, ideas for survey branding or wording for specific questions can be posted and input obtained.

This qualitative data can then help to shape the look and feel of any type of survey. Care should be taken to ensure that the input represents a broad spectrum of potential survey participants to prevent bias. It is also important to ensure that respondents understand that they are providing input on survey design and not commenting on the specific topics of the questions.

Identifying Respondent Groups

Social media foster connectedness—socially, professionally, geographically, even based on specific demographic characteristics or medical conditions. As a result, the social community provides a unique venue to find specific types of respondents. Operating from the premise that social media users are connected to other users with similar characteristics, researchers can build samples of a specific respondent type that they can then survey.

For example, working mothers often are connected to other working mothers through social media. If an appeal reaches one mother, she could be asked to recommend friends who are also working mothers to participate. This type of approach is referred to as "snowball sampling." According to Smart-survey's blog (2012) and Swicegood (2012), this approach can generate contacts to followers with specific characteristics to form a choice sample. These choice samples can be used to inform surveys where specific population groups are absent or form an online panel for further input on a particular transportation topic.

In 2011, researchers at the TTI used social media to reach bicycle riders in the Austin metropolitan area for a study. A project-related Twitter account (@CTracksAustin) invited local cyclists to download a smartphone application that would track their movements and provide researchers with detailed route information for planning purposes.

Using Social Media for Transportation Modeling

While researchers are working on ways that social media can be used to conduct statistically valid surveys, there is a great deal of ongoing work on the use of social media to supplement transportation surveys for purposes such as travel demand forecasting. Research investigating what Axhausen (2008) calls "the link between social networks, location choices, and travel" has a well-established history in transportation research, especially the connection of travel and information and communication technologies (ICT). Mokhtarian et al. (2006) and Sharmeen et al. (2010) provide overviews of numerous studies.

Traditionally, these connections have been assessed using face-to-face interviews and paper surveys. However, several challenges are associated with constructing social network data using these types of data collection, notably that respondents frequently have difficulty recalling their social network members (Carrasco et al. 2008). With the advent of online social networking sites (Facebook, LinkedIn, MySpace, etc.) over the past decade, much of this problem can be overcome. Users complete online profiles that often include demographic, educational, and geographic information that is shared with other users to create an online model of relationships and connections stored in a centralized database.

Changes to this online model of an individual's social network (e.g., a connection moves to a new city) are automated because the model is updated automatically when a connection changes information in his or her user profile. This can overcome an individual's cognitive difficulty in maintaining a large mental model of his or her social network that is subject to change over time.

Issues still exist when online social media information is used exclusively in surveys. Several of these issues were discussed earlier. In transportation and travel demand modeling, the strength of a social network connection is difficult to assess exclusively from online social media, as are other elements such as travel histories. For this reason, studies examining the connection between social networks and travel typically incorporate a respondent survey or interview element (Carrasco et al. 2008; Larsen, Axhausen, and Urry 2006).

A recent example of an attempt to use online social media in transportation modeling is an effort by researchers at the Georgia Institute of Technology. They are using social media to supplement a travel survey to better understand what has been noted as the "largest source of error" in travel demand modeling: destination choices (Zhao and Kockelman 2002 cited in Bernardin et al. 2009). Researchers collect individuals' social networking user profile information to supplement a traditional travel diary survey that is then used to investigate the link between an individuals' propensity for long-distance travel and the size and distribution of his or her online social network. The study focuses on Facebook, which has a built-in application programming interface (API) that allows external applications to interact with Facebook on behalf of users (Facebook 2013a).

Considering that Facebook user profiles include demographic, educational, and geographic information for "more than a billion monthly active users" (Facebook 2013b) across the globe, Facebook maintains a tremendous volume of data that could be of value in demonstrating how individuals choose travel destinations based on the size and geography of their social network.

The Georgia Tech study developed a two-part online survey. Participants are first asked for supplemental demographic data and a diary of long-distance air-travel trips taken in the previous year. Respondents then enter their Facebook login information. This triggers execution of a program that anonymously collects demographic and geographic data from the participant's Facebook profile and connections list to identify the size and geographical distribution of his or her social network. The collected social network data are then incorporated into a destination choice model and statistically analyzed to examine their effects on destination choice modeling accuracy.

Although data collection for this study is currently ongoing, the Georgia Tech team expects that incorporating social network data into destination choice modeling will yield statistically significant reductions in modeling errors. From a practical view, this research could directly benefit aviation or passenger-rail planners in identifying future routes and inform revenue managers who set prices for certain routes. On a conceptual basis, this work could introduce a powerful new tool for capturing online social network characteristics to supplement traditional survey-based transportation studies.

Conclusions

Transportation surveys tend to follow the guidelines and standards established in the general survey research industry. As a result, the use of social media in transportation surveys is largely limited to communications about surveys. Examples include posting informational videos to YouTube or promoting a link to a non-scientific survey. More recently, social media has been used to promote participation in online transportation communities where the use of web-based surveys is common.

Despite the limited use of social media for this purpose, survey researchers as a community, and transportation researchers in particular, are investigating how to better leverage the power of social media to improve survey participation and address statistical concerns regarding the need for representative samples to allow generalizations to be drawn from survey results. As research proceeds into these areas, it is anticipated that social media will become stronger tools for conducting transportation surveys. In the meantime, data mining and analysis methods are beginning to supplement traditional surveys for transportation planning analysis; some transportation-related uses are described in the next section.

CHECKING THE URBAN PULSE: SOCIAL MEDIA DATA ANALYTICS FOR TRANSPORTATION APPLICATIONS

Satish V. Ukkusuri, Samiul Hasan, and Xianyuan Zhan

The widespread use of social media provides extraordinary amounts of user-generated data every day. These data include extensive details and provide opportunities to transportation planners and engineers to tackle age-old transportation problems at spatial and temporal resolution levels that were previously unimaginable.

For instance, the availability of detailed geolocation information allows a transportation agency to visualize the "urban pulse" in real time, understand human behavior, and observe system performance with greater precision and coverage. Transportation analysis can exploit the vast amounts of data to accurately measure system performance and user behavior in real time. Similarly, although transportation agencies started using social media very recently, mainly to push out information about services and real-time updates, they have new opportunities to understand what their customers perceive in real time using social media data.

Overview

Similar to the sensors embedded in the physical environment, emerging social media systems can be viewed as a rapidly growing web of social sensors presenting tremendous potential to be used to observe the behaviors of millions of people in real time. These sensors will let researchers observe a diverse set of indicators ranging from consumer choices to public opinions about any issue from user movements in a city to public moods in different places and times.

Widely used social media platforms (e.g., Twitter, Facebook, and GooglePlus) publish exabytes of real-time user-generated data in the form of status updates, media sharing, and check-ins every day. The widespread use of smartphones equipped with GPS devices has made location-sharing services in social media more popular. Location-based services like Foursquare, Facebook Places, and Google Latitude create unprecedented access to extensive details about human activities and choices on a massive scale.

Combining this geospatial data with valuable information about social connections has the potential for detailed examinations of human mobility, activity participation behavior, and the broader connections among social networks, urban land use, and human movements. Two specific recent trends show that the data available from social media will be large scale, representative, and not temporary: (1) the availability of location-based services in many applications and (2) the large penetration of location-based services.

Location-Based Services in Social Media

With the advent of easily deployable location acquisition technologies such as GPS and Wi-Fi, location-based services have emerged in all forms of social media. These services enable people to attach location information to their existing online social networking activities in a variety of ways. The location attribute can provide useful information about daily activities and interactions with surrounding environments and social networks.

Examples of location-based services include systems by which users can upload geotagged photos to an online photo sharing system such as Flickr (Hollenstein and Purves 2010). They can send status messages in real time about an event from the place where the event is happening, for instance, via Twitter (Sakaki et al. 2010; Collins et al. 2012). Users can share their present activity locations on a social media channel such as Foursquare (Cheng et al. 2011; Noulas et al. 2012; Gundecha and Liu 2012) and share their social or group activities on Facebook Places (Chang and Sun 2011). It is possible to record travel routes with GPS trajectories to share travel experiences in an online community, for example, on GeoLife (Zheng et al. 2010) or log jogging and bicycle trails for sports analysis and experience sharing on Bikely (Counts and Smith 2007).

These location-based services enable researchers to collect information on various human activities at several geographic and temporal levels. Such information can help answer questions related to human behaviors based on real-world large-scale observations and aid development of interesting applications in many different fields.

Smartphone Growth
Market analysis shows that by 2013 the world had one billion smartphones (Arete 2012). A recent survey from the Pew Internet & American Life Project (Zickuhr 2012) found that smartphone ownership among American adults increased from 35% in 2011 to 46% in 2012. These smartphones are typically equipped with location acquisition technologies suitable for location-based services offered by social media applications.

The rapid growth of smartphone ownership led to corresponding increases in the use of location-based services. According to a comScore report (2011), nearly one in five smartphone users taps into check-in services like Facebook Places, Foursquare, and Gowalla. The Pew report also found that almost three quarters (74%) of smartphone users in 2012 (up from 55% in 2011) access real-time location-based information to find directions or recommendations.

The same report found as of February 2012 that 18% of smartphone owners used geosocial services of social media platforms like Foursquare, Gowalla, and Facebook Places, to "check in" to certain locations and share them with friends—an increase from the 12% noted in May 2011

(Zickuhr 2012). This means that 10% of all U.S. adults (18 and older) used smartphone check-in services in 2012, up from 4% in 2011. By 2013, more than 82 million users will subscribe to location-based social networking services (ABI Research 2008). Businesses are expected to spend $1.8 billion on location-based advertising by 2015 (ABI Research 2010).

This section explains how geolocation data from social media can rapidly transform the modeling of transportation system performance, determine user sentiments, and model land use characteristics on an urban scale that was not possible earlier. This section is intended to be a review of existing and potential tools that can be used for these purposes. Specific details of the models are omitted and can be found in other publications, including those of the authors.

This section will first discuss potential research questions using social media data for solving transportation-related problems and then present several examples of applications or models that used social media data. Finally, the issues and challenges of using social media data for transportation research will be presented.

Potential Research Problems Using Social Media Data

This section briefly describes a few interesting research problems that are relevant for transportation agencies and can be studied using social media data: (1) visualizing urban dynamics in real time, (2) understanding individual activity participation behaviors, (3) roles of communication patterns and social networks in activity choices, (4) uses of social traffic sensors, and (5) measuring user perceptions of services.

Real-Time Visualization of Urban Dynamics
Some social media users choose to share their locations when they visit restaurants, shopping malls, movie theaters, and similar venues via applications like Facebook Places or Foursquare. Users' locations are also available when they post status updates or media on their social networks from mobile devices. These data provide instantaneous geolocations and present tremendous potential for revealing the dynamics of urban environments and population activities in real time.

By overlaying individual location information on geographic references, the patterns of urban daily life can be characterized. This visualization of the real-time urban pulse will enable researchers to see how different places are used in the course of a day. Synthesizing this mobility data with traffic flow information, analysts can infer the relationship between distributions of traffic and people in urban spaces.

Understanding Individual Activity Participation and Location Choice Behaviors

Although the presentation of social media data on geographic spaces can provide interesting information, they can also provide further knowledge about human behaviors. In addition to attaching a geolocation, location-based services in social media enable people to share their activity-related choices (check-ins) in their virtual social networks (Facebook, Foursquare, Twitter, etc.). The large volume of geolocated human activity data offers the opportunity to analyze individual activity participation and location choice behaviors over a longer period.

When an individual gives a status in social media, information such as the coordinates of the place and the time of the status are recorded. Over a longer period, such information reveals an individual's longitudinal movement patterns. Understanding individual longitudinal location choices and the corresponding times of the day will contribute to the understanding of spatiotemporal patterns of human movements from one place to another. Most importantly, this information allows large-scale analysis of human mobility behavior and comparisons of activity behaviors of different urban areas on a global level.

Influence of Communication Patterns in Social Media and Social Networks on Activity Participation

Communication tools such as online social media have changed the ways people exchange information and experiences. Individuals can share their experiences about locations they visit along with the routes and modes they used to get there. The online tools thus act as communication channels to transmit information from one individual to another.

Applications like Waze can show patterns of information sharing and individual reactions to this information. Another benefit is that such information can help individuals to optimize their daily activity travel plans. An important question to explore is how these communication patterns within social networks influence collective activity and travel behaviors.

In the past few years, researchers of activity-based modeling (Carrasco et al. 2008; Habib et al. 2008) studied activity scheduling processes that depend on individual characteristics and also on people with whom an individual interacts. Specifically, an individual's social network plays an important role in his or her social activity and travel behavior. Researchers have developed advanced behavioral models to capture this social dimension of behavior based on questionnaire surveys. Longitudinal data on user behavior from social media can lead analysts to develop more advanced models capturing the dynamics of social networks and their associated influence on individual activity and travel behavior.

Social Traffic Sensors

Applications like Waze created opportunities to collect and disseminate real-time traffic information at almost no cost. What is missing now is a traffic assignment model that can use the information provided by users to suggest an optimal path based on future traffic flow. The challenge, however, will be to incorporate connectivity of the social network, users' communication patterns, and actual traffic flow. Thus an online dynamic traffic assignment (DTA) model can be neatly integrated with this type of social networking application. Figure 5.4 depicts the concept.

Measuring User Perceptions of Services

An important aspect of social media is that users can individually or collectively express their opinions, champion a cause, or call for action on a

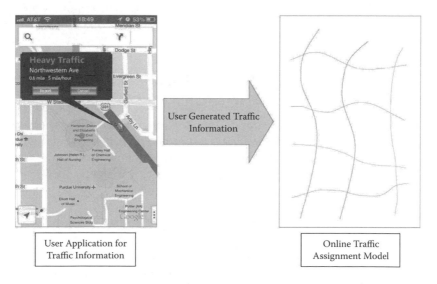

Figure 5.4 Online traffic assignment model incorporating user-generated information content. (From: www.maps.google.com. Google Maps © 2013. With permission.)

191

subject instantaneously. This has been demonstrated many times recently, most notably by the Arab Spring that led to a political reshaping of the Arab world (Howard and Hussain 2011).

Researchers are beginning to "mine" these opinions from social media outlets to analyze general public perceptions or sentiments on many subjects. Sentiment analysis or opinion mining is the examining of a text through learning tools to understand the positive or negative connotation surrounding it. Sentiment analysis can assist companies to determine how a brand is perceived in relation to value and quality. Subtasks of sentiment analysis include determining subjectivity, polarity (positive or negative), degree of polarity, and classification of subject and author.

Similarly, sentiment analysis based on data gathered from social media may change the way transportation organizations measure user satisfaction. In situations that require proactive responses, carefully analyzed data from social media can serve as prognostic tools to provide timely and effective transit services. Online social media allows commuters to voice their concerns about current service, safety, sanitary conditions, and the like in real time. Until now, transportation agencies had few methods other than surveys and focus groups for gauging customer satisfaction. These methods were not always practical and often their results were not timely. Social media data can provide useful and immediate information for measuring rider perception.

Illustrative Modeling Frameworks of Social Media Applications

Based on the above background, this section will describe potential approaches for modeling the various issues cited: (1) using social media data to explain urban human mobility patterns, (2) modeling individual-level activity patterns, (3) measuring transit riders' satisfaction levels, and (4) inferring urban land use patterns.

Explaining Urban Human Mobility Patterns

Recently researchers have characterized human mobility patterns using distance-based measures based on alternative datasets collected from mobile phones (González et al. 2008), bank note movements (Brockmann et al. 2006), and subway smart card transactions (Hasan et al. 2012a). These studies, however, do not fully reveal the relationship between the selections of destinations for different purposes and mobility dynamics without analyzing the purposes behind these human movements. By including spatiotemporal dimensions and activity categories (purposes) into analyses, analysts can discover more realistic and detailed descriptions of human mobility dynamics.

Researchers at Purdue University (Hasan et al. 2013) used a large dataset of Twitter posts to analyze urban human mobility patterns. Twitter users can post short messages known as tweets that are attached with the corresponding geolocations when users grant permission. In addition to posting status messages, Twitter allows its users to post statuses from third-party check-in services (e.g., Foursquare). The researchers used a large-scale check-in dataset available from another study (Cheng et al. 2011). The dataset contained check-ins in New York City from February 25, 2010, through January 20, 2011, with an average 25 check-ins per user.

One of the major advantages of using these data was the ability to identify activity purposes. Each check-in observation reports a short link to the original check-in source agency (e.g., Foursquare) that gives information about the category of the visited location when queried. Based on the types of visited locations, check-ins can be classified into different activity categories.

In their work (Hasan et al. 2013), the authors investigated the characterization and visualization of aggregate human mobility and activity patterns by constructing a virtual grid reference of a New York City map into square cells of 200 by 200 m. Spatial distributions of visits to different places were determined for various activity purposes by counting the number of purpose-specific visits within each cell (see Figure 5.5) and computing the proportion of visits to each cell for each activity category. This generated activity distribution map shows the popular places within the city and the functionality of each part of the urban area. These check-in distributions differ based on activity category. This suggests a strong influence of urban context on destination choices.

Using non-parametric density estimation methods, time-dependent activity density maps were constructed. This approach allows researchers to also visualize different activities in a city and thus capture the pulse of urban human activities.

This kind of spatiotemporal analysis of mobility dynamics can lead to a mobility model that plays an important role for building more realistic agent-based simulation tools for contexts in which mobility is a critical factor. For instance, researchers have long been interested in building agent-based simulation models to track disease spread in cities. A simple mobility model reproducing human movements in the city can provide

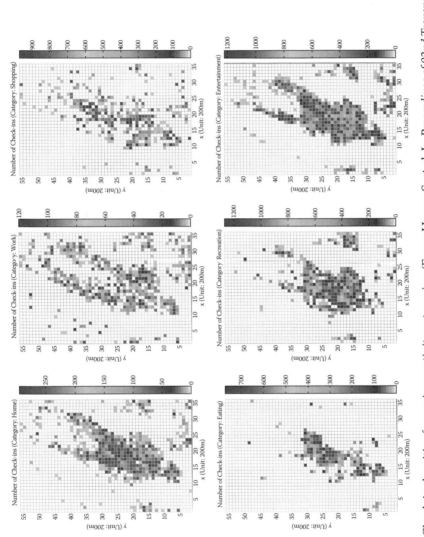

Figure 5.5 Check-in densities for various activity categories. (From: Hasan, S. et al. In *Proceedings of 92nd Transportation Research Board Meeting*. With permission.)

the spatiotemporal details of the movements that are input to the disease spread model. Hasan et al. (2013) proposed a mobility model with a simple location choice process that satisfies the general principles of human movements.

Modeling Individual Activity Patterns

With the availability of new data sources from mobile phones and GPS devices, interest in modeling individual activities over time has been renewed (Axhausen et al. 2002; Ettema and van der Lippe 2009; Arentze et al. 2011). Using the longitudinal geolocation data available from social media, it is now possible to develop complex models inferring individual activity routines.

For instance, Hasan and Ukkusuri (2012b) propose new methodologies for modeling individual activity patterns using location-based data from smartphone check-in services. These methodologies are flexible enough to use various types of data, for instance, geolocation information collected on a large scale by recording the smartphone GPS coordinates.

The authors propose a data-driven modeling technique to extract individual-level weekly activity patterns from these geolocation data. The type of activity is considered as the basic unit of data to be picked from a set of activities. A user completes a collection of activities for a given number of weeks and the sequence of activity participations represents a routine.

The activity pattern inference problem can be defined as follows. Based on the activity labels for a number of weeks over which an individual participates in various activities, how can researchers determine the hidden activity patterns where each activity pattern represents a distribution of activities.

The activity pattern model starts by choosing a distribution over routines for a given week. Based on the distribution of routines for a week, activity labels are generated by assigning a routine. The output from the model is a vector of activities for a week. Several meaningful weekly activity patterns are found from individual-shared activities on social media. The discovered patterns mainly capture participation in non-work-related flexible activities.

The model can be extended to capture user-specific patterns capable of identifying specific users who contribute to a specific pattern. An extension to this model is proposed to account for the missing activities—one of the major limitations of online geolocation data. This approach demonstrates that with additional information the model can predict the missing activities in individual activity behavior. The geolocation data can then be used to supplement surveys that inform activity-based models of travel behavior utilized by many metropolitan planning agencies.

Sentiment Analysis of Social Media Data for Measuring Transit Rider Satisfaction

With the advent of social media, people can express and share their opinions or perceptions on a variety of subjects instantaneously. Researchers have been analyzing these opinions from social media outlets (i.e., Twitter, Facebook) to measure public perceptions on various issues. By analyzing the sentiments of the status messages posted on social media, transit agencies can measure rider satisfaction regularly in a cost-effective fashion.

In earlier work that includes the authors of this paper, Collins et al. (2012) demonstrated how social media data can be used for measuring rider satisfaction with the Chicago Transit Authority's rail system. Popularly known as the L, the system has eight separate lines denoted by different colors. The research team collected Twitter status updates or tweets that contained the keywords of all combinations of L train names near the city. For example, when looking for tweets relative to the Blue Line L, the search command was "Blue Line near:Chicago." This method of manual searching (line color:location) was used to collect tweets for all the lines.

The team used SentiStrength (Thelwall et al. 2010), a machine learning program that detects the sentiment value of a short text, for analyzing the sentiments of the riders' tweets about the transit system. SentiStrength uses machine learning for analysis to determine the average negative and positive sentiments or the minimum negative or maximum positive sentiments of a collection of texts.

The core of the algorithm is the sentiment word strength list (Thelwall et al. 2010). The list can be updated with new words relevant to a specific topic, allowing the algorithm to better judge sentiment polarity. Several sample tweets and the corresponding sentiment strengths are shown in Examples 1 through 3. When assigning a sentiment value to a tweet, the maximum positive and negative values of all the words in the tweet are assigned positive and negative strengths. Each tweet is assigned a minimum negative strength (–1) and a minimum positive strength (+1) in the absence of negative and positive words within the tweet.

Example 1

I	hate	the	blue	Line.	Positive	Negative
0	–2	0	0	0	1	–2

Example 2

Power	At	green	line	station	out.	Eerie!	Positive	Negative
0	0	0	0	0	0	−2	1	−2

Example 3

The	blue	line	may	not	be	the	most	glamorous	mode	of	transportation,
0	0	0	0	0	0	0	0	2	0	0	0

but	by	golly	I	don't	envy	the	people	sitting	in	traffic	right	now.	Positive	Negative
0	0	0	0	0	3	0	0	0	0	0	0	0	3	−1

SentiStrength incorporated algorithms to help correct tweeting "norms" by fine tuning the sentiment strengths using a set of training data. To optimize the sentiment word strengths, the training algorithm uses baseline human-allocated term strengths to assess whether an increase or decrease in the term strength of the word strength list would increase the accuracy of the classifications. To make the algorithm compatible with the tweeting norm of the L riders, the authors manually classified 500 tweets assigning positive or negative weights to each word in the tweets to optimize the sentiment strengths of the dictionary words. Table 5.1 shows an example of this process.

The total positive and negative sentiments were analyzed to identify a significant increase or decrease over time. If significant changes are found, the time period can be scrutinized further to determine contributing factors to the change in sentiment. Average sentiments were computed to calculate an average overall sentiment. It is important to characterize the overall experience of all riders for a particular period. The benefit of using an average sentiment was to ensure that one or two tweets alone did not greatly affect the satisfaction level for a time period. Standard error was also calculated to determine the error from the mean during a specific period.

Analysis showed that transit riders are more inclined to assert negative sentiments about a situation than positive sentiments (Figure 5.6). The findings indicate that sentiment analysis can successfully detect transit system rider sentiments in real time. An online system can measure performance from user status feeds on social media in real time. Such a system can provide valuable feedback to transit managers and be used

Table 5.1 Sample Word List

Word	Sentiment Range (–5 to –1 and +1 to +5)		
	Previous Value	**Change**	**Current Value**
Ugh	–3	+1	–2
Hate	–3	+1	–2
Heavenly	4	–1	3
Obnoxious	–3	+1	–2
Petty	–1	–1	–2
Scary	–2	–1	–3
Screw	–3	+1	–2
Well	2	–1	+1
Worse	–3	+1	–2
Worst	–3	+1	–2
Wtf	–3	+1	–2
Yay	2	–1	+1

to improve services. Most importantly, riders' dissatisfaction about incidents can be quantified through sentiment analysis of social media data, allowing an agency to understand what types of incidents change rider perceptions the most.

Inferring Urban Land Use Patterns

Urban land use is the key term in the language of urban planning. The traditional land use identification approach involves on-site investigations, surveys, and questionnaires. All these measures require intensive labor resources, time, and money. The mutual influence of human behavior (or human mobility and activity patterns) and land use classification, especially inferring land use based on activity and mobility patterns, has attracted increasing interest in the recent literature.

Various approaches using novel data sources such as mobile phone data, taxi GPS data, and point-of-interest data have been explored (Soto and Frias-Martinez 2011a, 2011b; Qi et al. 2011; Toole et al. 2012; Yuan et al. 2012). However, all these approaches have certain limitations, such as

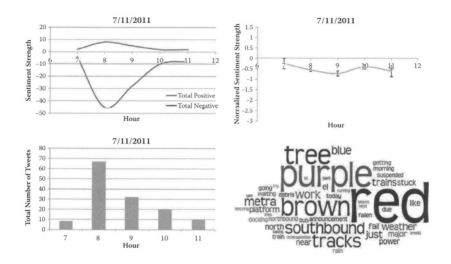

Figure 5.6 Total sentiment strength, normalized sentiment strength, total number of tweets, and tag cloud at 8:00 a.m. The tag cloud contains words that provide details about trains (red, purple, brown), reasons for incidents (tree, weather, fail), and general mood (yelling). (From: Collins, C. et al. 2012. In *Proceedings of 91st Transportation Research Board Meeting*. With permission.)

low recognition accuracy and limited land use type identification, mainly because of insufficient detail in the data.

Zhan (2012) developed a land use inference method by utilizing social media check-in data. The advantages of social media check-in data compared to other data sources lie in the wide coverage range, easy and convenient accessibility, and the invaluable information based on activity participation. Zhan conducted a case study in New York City, using a dataset of more than 460,000 tweets with Foursquare check-ins from 18,440 users. The raw check-in data contained only latitude–longitude coordinates and the activity category. The number of check-ins of each category for each time period (eight periods per day) and each day was counted for each cell. The map was divided into small cells of a given size. The data were first aggregated by weekdays and weekends, and information from six activity categories (home, work, dining, recreation, shopping, and social service) was considered.

In a cluster analysis, Zhan (2012) developed an algorithm to classify the urban area into land use types. Based on an actual land use zoning map provided by the New York City Department of City Planning, the four land use types were classified as residential, commercial, open

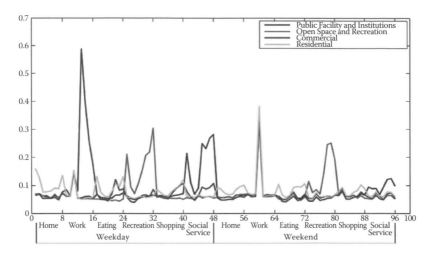

Figure 5.7 Clustering centers for identified land use types; y-axis = normalized activity intensity. (From: Zhan, X. 2012. MSE Thesis, Purdue University. With permission.)

space and recreation, and public facilities and institutions. The clustering centers and spatial representation of the clustering results are shown in Figure 5.7. Due to limitations in the data, the model could not infer certain land uses such as industrial or manufacturing. The overall results from the 200 m × 200 m grids showed a close match with the actual land use zoning situation (Figure 5.8), suggesting the proposed method is a good candidate for land use inference.

Challenges of Using Social Media Data

Location-based data from social media are available in large sample sizes for long periods (even for a year) without significant costs and provide exact locations and timing information about events and individual activity participation. However, these data have some issues that restrict their use for existing transportation modeling techniques: (1) concerns about the privacy of the social media users, (2) the impacts of self-selected social media users and their biases toward reporting certain types of activities, and (3) lack of detailed information on activity participation and sociodemographic characteristics. Appropriate methodologies addressing these limitations are required to use these types of data.

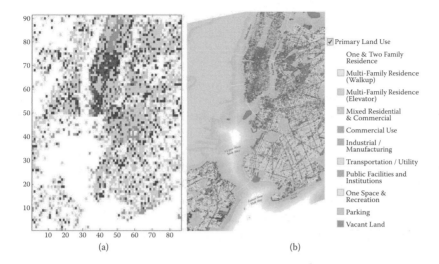

Figure 5.8 Empirical land use inference validation. (a) Inferred land use pattern. (b) Actual land use zoning map provided by New York City Department of City Planning. (From: Zhan, X. 2012. MSE Thesis, Purdue University. With permission.)

Privacy Concerns

The pervasive uses of location-based services in social media pose a new threat to users' security and privacy concerns. With more location-based information available, users are more vulnerable to malicious activities. As these data provide crucial information about the patterns of user movements and activities without significant barriers, protection of these data is very important for individual security.

While users can restrict their information through privacy measures available in applications, social media authorities should be more careful when publishing these data. Maintaining the anonymity of users will be a big challenge for researchers who utilize social media data. Researchers and public agencies working with this enormous amount of data should adopt secure storage and communication protocols to protect the identities of the users.

Selectivity Biases

A critical limitation of social media data is the lack of representativeness of samples. A survey conducted by comScore (2011) found that U.S. users of smartphones and location-based services are slightly over-represented by younger people. Researchers should be cautious about this bias when using results from analyzing social media data. However, it is expected that such biases will decrease with wider use of these services in the future. The equity of social media is discussed in Chapter 6.

Missing Information

Although social media data allow huge coverage for longer time periods, they lack much of the detail required for traditional modeling method-ologies. For instance, user information from social media does not reveal socioeconomic characteristics and activity sharing does not indicate the start and end times or durations of activities. Modeling methodolo-gies should look for novel techniques that can either infer this missing information (e.g., user socioeconomic characteristics) or not require such information for achieving the required objectives. Modeling techniques dealing with hidden variables will offer some advantages in this regard.

Conclusions

The rapid growth of smartphones and the introduction of location-based services in social media enable researchers to collect user-generated data with unprecedented details and coverage. These vast amounts of data contain valuable hidden information about large populations. However, to make meaningful inferences from these data, researchers require novel techniques to account for their large scales and limitations. The advances in machine learning techniques capable of handling large-scale datasets show a promising future for analyzing social media data containing rel-evant and continuous information about millions of users.

Social media data can be used to analyze urban mobility patterns and individual activity patterns, measure transit rider satisfaction, and infer urban land use. Assuming that the challenges of privacy, selectivity bias, and missing information can be overcome, transportation agencies and researchers should avail themselves of these opportunities offered by social media data to develop new methodologies and build applications solving many real-world problems.

REFERENCES

ABI Research. 2010. Mobile advertisers forecast to spend $1.8 billion on location-based campaigns in 2015.

ABI Research. 2008. Eighty-two million location-based mobile social networking subscriptions by 2013.

Arentze, T.A., D. Ettema, and H.J.P. Timmermans. 2011. Estimating a model of dynamic activity generation based on one-day observations: method and results. *Transportation Research B*, 45, 447–460.

Arete. 2012. Getting to a billion smartphones in 2013. http://www.telco2research. com/articles/AN_Arete-billion-smartphones-2013_Summary

Axhausen, K.W. 2008. Social networks, mobility biographies, and travel: survey challenges. *Environment and Planning B*, 35, 981–996.

Axhausen, K.W., A. Zimmermann, S. Schonfelder et al. 2002. Observing the rhythms of daily life: a six-week travel diary. *Transportation*, 29, 95–124.

Bassett, D., J. Beard, J. Ewing et al. 2012. Conducting surveys using social media. http://prezi.com/xn0bnsg7sxvu/conducting-surveys-using-social-media/

Behavior Inventory (TBI), http://www.youtube.com/watch?feature=player _embedded&v=bALKrAjaX3Q0

Bernardin, V., F. Koppelman, and D. Boyce. 2009. Enhanced destination choice models: incorporating agglomeration related to trip chaining while controlling for spatial competition. Ohio Travel Demand Model Users Group.

Boyne, G.A. 2003. Sources of public service improvement: a critical review and research agenda. *Journal of Public Administration Research and Theory*, 13, 367–394.

Brabham, D.C. 2009. Crowdsourcing the public participation process for planning projects. *Planning Theory*, 8, 242–262.

Bradwell, P. and S. Marr. 2008. Making the Most of Collaboration an International Survey of Public Service Co-Design: DEMOS Report 23. DEMOS, in association with PricewaterhouseCoopers (PWC).

Brockmann, D., L. Hufnagel, and T. Geisel. 2006. The scaling laws of human travel. *Nature* 439, 462.

Burke, J.A., D. Estrin, M. Hansen, A. Parker, N. Ramanathan, S. Reddy, and M.B. Srivastava. 2006. Participatory Sensing. Proc. of 1st Workshop on Wireless Sensor Web, WSW'06, Boulder, October 31, pp. 1–5.

Carrasco, J.A., B. Hogan, B. Wellman et al. 2008. Collecting social network data to study social activity–travel behavior: An egocentric approach. *Environment and Planning B*, 35, 961.

Chang, J. and E. Sun. 2011. Location 3: how users share and respond to location-based data on social networking sites. In *Proceedings of AAAI Conference on Weblogs and Social Media*, pp. 74–80.

Cheng, Z., J. Caverlee, K. Lee et al. 2011. Exploring millions of footprints in location sharing services. Fifth International AAAI Conference on Weblogs and Social Media.

Chiao, H. et al. 2011. Continuous improvement in regional household travel surveys: the NYMTC Experience. Annual Meeting of Transportation Research Board.

Collins, C., S. Hasan, and S. V. Ukkusuri. 2012. A novel transit rider satisfaction metric: rider sentiment measured from online social media. In *Proceedings of 91st Transportation Research Board Meeting*.

comScore. 2011. Nearly 1 in 5 smartphone owners access check-in services via their mobile device. http://www.comscore.com/Press_Events/Press_Releases/2011/5/Nearly_1_in_5_Smartphone_Owners_Access_Check-In_Services_Via_their_Mobile_Device

Counts, S. and M. Smith. 2007. Where were we? Communities for sharing space-time trails. In *Proceedings of 15th Annual ACM International Symposium on Advances in Geographic Information Systems*, pp. 10:1–10:8.

DeVall, R. Undated. Social media versus online panel sampling: not all respondents are created equal. *GreenBook*. http://www.greenbook.org/marketing-research.cfm/social-media-vs-online-sampling-not-all-respondents-created-equal-03914

Edison Research. 2012. The social habit. http://www.edisonresearch.com/wp-content/uploads/2012/06/The-Social-Habit-2012-by-Edison-Research.pdf

Ettema, D., T. Arentze, and H. Timmermans. 2007. Social influences on household location, mobility and activity choice in integrated micro-simulation models. CD-ROM. Paper presented at Workshop Frontiers in Transportation, Amsterdam.

Ettema, D. and T. van der Lippe. 2009. Weekly rhythms in task and time allocation of households. *Transportation*, 36, 113–129.

Facebook. 2013a. Facebook developers: core concepts. http://developers.facebook.com/docs/coreconcepts/

Facebook. 2013b. Key facts. http://newsroom.fb.com/Key-Facts

Goldstein, S.M., R. Johnston, J.A. Duffy et al. 2002. The service concept: missing link in service design research? *Journal of Operations Management*, 20, 121–134.

González, M.C., C.A. Hidalgo, A. L. Barabási. 2008. Understanding individual human mobility patterns. *Nature*, 453, 779–782.

Gundecha, P. and H. Liu. 2012. Mining social media: A brief introduction. *INFORMS TutORials in Operations Research*, 9, 1–17.

Habib, K.M., J.A. Carrasco, and E.J. Miller. 2008. Social context of activity scheduling: discrete–continuous model of relationship between "with whom" and episode start time and duration. *Transportation Research Record*. 2076, 81–87.

Hasan, S., X. Zhan, and S.V. Ukkusuri. 2013. Understanding urban human activity and mobility patterns using large-scale location-based data from online social media. In *Proceedings of 92nd Transportation Research Board Meeting*.

Hasan, S., C.M. Schneider, S.V. Ukkusuri et al. 2012a. Spatiotemporal patterns of urban human mobility. *Journal of Statistical Physics*.

Hasan, S. and S.V. Ukkusuri. 2012b. Modeling individual activity patterns using large-scale geolocation data. Working Paper. Purdue University.

Henning, J. 2010. Surveys and social media. *Listening Post*, April 8. http://blog.vovici.com/blog/bid/27221/Surveys-Social-Media (accessed January 8, 2013).

Hollenstein, L. and R. Purves. 2010. Exploring place through user-generated content: using Flickr to describe city cores. *Journal of Spatial Information Science*, 1, 21–48.

Howard, P.N. and M.M. Hussain. 2011. The role of digital media. *Journal of Democracy*, 22, 35–48.

Howe, J. 2006. The rise of crowdsourcing. *Wired*. http://www.wired.com/wired/archive/14.06/crowds_pr.html

King, S.F. and P. Brown. 2007. Fix my street or else: Using the Internet to voice local public service concerns. In *Proceedings of ACM Conference on Theory and Practice of Electronic Governance*, pp. 72–80.

Larsen, J., K.W. Axhausen, and J. Urry. 2006. Geographies of Social Networks: Meetings, Travel and Communications. *Mobilities*, 1(2), 261–283.

Machado, S., R. Jose, and A. Moreira. 2012. Social interactions around public transportation. In *Proceedings of Seventh Iberian Conference on Information Systems and Technologies*, pp. 1–6 and 20–23.

Meagor, D. n.d. Article dashboard: how social media can help surveys. http://www.articledashboard.com/Article/How-Social-Media-Can-Help-Surveys/2171132

Mokhtarian, P.L., I. Salomon, and S.L. Handy. 2006. The impacts of ICT on leisure activities and travel: a conceptual exploration. *Transportation*, 33, 263–289.

North County Transit District, https://twitter.com/GoNCTD/status/10895339088

Noulas, A., S. Scellato, R. Lambiotte, M. Pontil, and C. Mascolo. 2012. A tale of many cities: universal patterns in human urban mobility. *PLoS ONE*. 7.

Parker, D.J. 2008. Transit cooperative research program synthesis. In *AVL Systems for Bus Transit: Update*. Washington: National Academy Press.

Potomac and Rappahannock Transportation Commission, https://www.facebook.com/notes/potomac-and-rappahannock-transportation-commission/take-the-teen-summer-pass-survey-and-you-could-win-a-100-visa-gift-card/10151147988089883

Poynter, R. 2010. *Handbook of Online and Social Media Research*. New York: John Wiley & Sons.

Prahalad, C.K. and V. Ramaswamy. 2000. Co-opting customer competence. *Harvard Business Review*, 78, 79–90.

Qi, G., X. Li, S. Li et al. 2011.Measuring social functions of city regions from large-scale taxi behaviors. In *Proceedings of IEEE International Conference on Pervasive Computing and Communications Workshops*, pp. 384–388.

Rose, J. 2011. How's our tweeting? TriMet asks readers to take Twitter poll. *The Oregonian*. http://blog.oregonlive.com/commuting/2011/06/hows_are_tweeting_trimet_asks.html

Saad-Sulonen, J., A. Botero, and K. Kuutti. 2012. A long-term strategy for designing (in) the wild: lessons from the urban mediator and traffic planning in Helsinki. In *Proceedings of ACM DIS*, pp. 166–175.

Sakaki, T., M. Okazaki, and Y. Matsuo. 2010. Earthquake shakes Twitter users: real-time event detection by social sensors. In *Proceedings of 19th International Conference on the Worldwide Web*.

Schwartzel, E. 2010. Port Authority takes to Twitter to report problems. *Pittsburgh Post-Gazette*.

SEPTA. 2011. https://www.facebook.com/notes/southeastern-pennsylvania-transportation-authority-septa/survey-for-future-planning/210691082280 535

Sharmeen, F., T. Arentze, and H. Timmermans. 2010. Modelling the dynamics between social networks and activity–travel behavior. Paper presented at 12th World Conference on Transport Research.

Silvertown, J. 2009. A new dawn for citizen science. *Trends in Ecology & Evolution*, 24, 467–471.

Sit, M. 2012. Like what we do? Don't like it? Tell us. *Write On Metro.* http://blogs.ridemetro.org/blogs/write_on/archive/2012/11/30/Like-What-We-Do_3F00_-Don_2700_t-Like-It_3F00_-Tell-Us.aspx

Smart-survey blog. 2012. The benefits of conducting surveys through social network sites. http://blog.smart-survey.co.uk/the-benefits-of-conducting-surveys-through-social-network-sites/

Soto, V. and E. Frias-Martinez. 2011a. Automated land use identification using cell-phone records. In *Proceedings of Third ACM International Workshop on MobiArch–HotPlanet.*

Soto, V. and E. Frias-Martinez. 2011b. Robust land use characterization of urban landscapes using cell phone data. Workshop on Pervasive Urban Applications, Ninth International Conference on Pervasive Computing.

Steinfeld, A., J. Zimmerman, A. Tomasic et al. 2012a. Mobile transit rider information via universal design and crowdsourcing. *Transportation Research Record*, 2217, 95–102.

Steinfeld, A., S.L. Rao, A. Tran et al. 2012b. Co-producing value through public transit information services. International Conference on Human Side of Service Engineering.

Steinfeld, A., R. Aziz, L. Von Dehsen et al. 2010. The value and acceptance of citizen science to promote transit accessibility. *Journal of Technology and Disability*, 22, 73–81.

Swicegood, J. 2012. Thinking outside the traditional survey research toolbox. SAPOR. https://blogs.rti.org/surveypost/2012/11/01/sapor-2012-thinking-outside-of-the-traditional-survey-research-toolbox/

Thelwall, M., K. Buckley, G. Paltoglou et al. 2010. Sentiment strength detection in short informal text. *Journal of the American Society for Information Science and Technology*, 61, 2544–2558.

Toole, J.L., M. Ulm, D. Bauer et al. 2012. Inferring land use from mobile phone activity. In *Proceedings of ACM SIGKDD International Workshop on Urban Computing.*

Turner, D.S., W.A. Evans, B. Wolshon et al. 2010. Transportation-oriented communication with vulnerable populations during major emergencies. Transportation Research Board Annual Meeting.

Von Ahn, L., B. Maurer, C. McMillen et al. 2008. Recaptcha: human-based character recognition via web security measures. *Science*, 321, 5895.

Wallace, J. 2012. MnDOT's online customer community. Transportation Research Board Annual Meeting. http://www.travelsurveymethods.org/pdfs/trb-workshop/2012/MNDOT%20OLC%20TRB%20Presentation%20January%2020%2012.pdf

Yoo, D., J. Zimmerman, and T. Hirsch. 2013. Probing bus stop for insights on transit co-design. In *Proceedings of Conference on Human Factors in Computing Systems.*

Yoo, D., J. Zimmerman, A. Steinfeld et al. 2010. Understanding the space for co-design in riders' interactions with a transit service. In *Proceedings of Conference on Human Factors in Computing Systems*, pp. 1797–1806.

Yu, S. and S. Kak. nd. A survey of prediction using social media. http://arxiv.org/ftp/arxiv/papers/1203/1203.1647.pdf

Yuan, J., Y. Zheng, and X. Xie. 2012. Discovering regions of different functions in a city using human mobility and POIs. In *Proceedings of KDD Conference.*

Zhan, X. 2012. Understanding the aggregate level urban activity patterns using large-scale geolocation data. MSE Thesis, Purdue University.

Zhao, Y. and K.M. Kockelman. 2002. The propagation of uncertainty through travel demand models: an exploratory analysis. *Annals of Regional Science,* 36, 145–163.

Zheng, Y., X. Xie, and W. Y. Ma. 2010. Geolife: a collaborative social networking service among user, location and trajectory. *IEEE Data Engineering Bulletin,* 33, 32–39.

Zickuhr, K. 2012. Three-quarters of smartphone owners use location-based services. Pew Internet & American Life Project.

Zimmerman, J., A. Tomasic, C. Garrod et al. 2011. Field trial of Tiramisu: crowd-sourcing bus arrival times to spur co-design. In *Proceedings of Conference on Human Factors in Computing Systems.*

6

Agency Considerations and Policies

Getting started with social media is the easy part. In fact, many government agencies jumped (or were pushed) into the social pool before they could sort out what to do and how to do it. Transportation agencies can set up one or more accounts and start posting, tweeting, pinning, or otherwise sharing information and conversing with citizens quickly. The real challenges for agencies start after the initial post. Commonly stated concerns include staffing requirements, ensuring digital access for all citizens, record-keeping protocols, and responding to online criticism. Many agencies also want guidance on how to evaluate their social media activities so they can communicate effectively and justify their efforts to their bosses and boards. This chapter focuses on such institutional considerations.

Starting off the chapter, Susan Bregman and Sarah Kaufman team up to write about protocols and policies that agencies can put in place to manage their social media activities. Agencies are often concerned that a social media presence will expose them to criticism, and Susan Bregman writes about approaches that agencies can take to recognize online threats and defuse them. Next, Katharine Hunter-Zaworski, Kari Watkins, and Sarah Windmiller write about information equity and strategies for ensuring that all citizens have access to the information posted on social media sites regardless of age, socioeconomic status, language, or disability. Finally, Eric Rabe and Susit Dhakal focus on tools to help transportation agencies determine return on investment for their social media activities.

209

WHAT'S THE WORST THAT CAN HAPPEN?
SOCIAL MEDIA PROTOCOLS AND POLICIES
Susan Bregman and Sarah M. Kaufman

As the preceding chapters demonstrate, social media offers multiple benefits to transportation organizations including dissemination of real-time information, crisis communications, marketing, feedback, and customer research. However, social media brings challenges as well, and some agencies have expressed concerns about the institutional requirements and potential pitfalls associated with their use. Organizational considerations for government agencies—especially those new to social media—can include:

- Which social media platform is the best choice for the agency?
- How much time is required to manage social media accounts?
- Which agency employees should have access to social media?
- Should any sites be off limits?
- When should the agency post updates?
- Does using social media put the agency at risk for viruses and malware?
- Are tweets part of the public record?

To address questions like these, many government agencies adopted policies and procedures to help manage their social media activities. This section identifies common concerns about agency use of social media and then presents examples from social media policies and best practices at transportation agencies.

What Is a Social Media Policy?

A social media policy is a set of guidelines and expectations for appropriate behavior when employees post content on social networking sites, either in the course of their jobs or on their own. While the practice is not universal, a growing number of agencies on all levels of government are adopting some form of social media guidance. In a 2012 survey of state departments of transportation conducted by the American Association of State Highway and Transportation Officials, 67% of respondents reported having an agency social media policy in place (AASHTO 2012).

Social media policies need not be stand-alone documents drafted from scratch. The Society for New Communications Research (2007) recommends starting with legacy guidelines including existing human

210

resources and communications policies. Dallas Area Rapid Transit (DART) took this approach. DART has no social media policy per se and created an addendum to its existing communications guidelines to address social media activities (Bregman 2012).

To help public sector organizations better understand how to manage social media activities, researchers at the Center for Technology in Government reviewed 26 government social media policies and interviewed 32 practitioners to identify common elements and best practices (Hrdinová et al. 2010). They determined that government social media policies typically incorporated guidance in eight areas: account management, employee access, employee conduct, acceptable use, content management, cyber security, legal issues, and user conduct.

Approaching social media policies from another angle, the Rudin Center for Transportation at New York University defined three broad goals for transportation agency social media activities: inform, motivate, and engage their followers. To accomplish these goals, agencies are encouraged to be accessible, informative, engaging, and responsive (Kaufman 2012b). The work of these researchers can be organized into four key elements:

- How to get started using social media
- How to manage limited staff resources
- How to manage employee use of social media
- How to address legal and institutional requirements

The remainder of this section reviews how transportation agencies developed formal or informal policies to address these commonly stated concerns about using social media.

Getting Started

The list of social media platforms keeps changing. Today's go-to channel may be tomorrow's afterthought. As discussed in Chapter 2, social media tools have different strengths and weaknesses and should be used accordingly. Transportation agencies are well served by matching the right tools to their communication and marketing goals. Some agencies may use only a Facebook page, while others always try the next best thing. The following are common ways that agencies use certain platforms:

- Twitter – Distribute short, focused, and time-sensitive messages.
- Facebook – Engage users and encourage sharing with longer informative posts.
- YouTube – Engage and inform customers through entertaining and/or instructional videos.
- Flickr – Share photos of agency news and events with the public and the media.
- Blog – Publish longer and more detailed updates about agency policies and events.

There is no right or wrong approach. The final decision for an agency likely depends on a combination of goals, staff skills, and resource availability. Figure 6.1 shows how an agency can convey information about the same incident on different social media platforms.

No matter how many social media channels an agency uses, it is important to ensure that people know about them. Many agencies include links

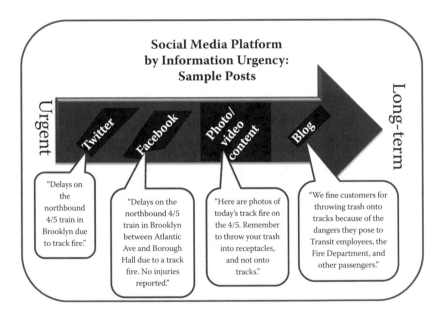

Figure 6.1 How a transit agency can craft a message about the same incident for different social media platforms. (From: Kaufman, S.M. 2012b. How Social Media Move New York. Part 2: Recommended Social Media Policy for Transportation Providers. Rudin Center for Transportation, New York University's Wagner School of Public Service. With permission.)

Figure 6.2 Texas Department of Transportation provides links to its social media accounts on its website. (Screen capture from Texas DOT website. With permission.)

to their various accounts on the home pages of their websites. Agencies with multiple accounts may direct users to an internal page showing the relevant links. Figure 6.2 shows a detail from a web page listing social media channels for the Texas Department of Transportation.

Content Management

Above all, social media requires frequent updates, and some social media policies and guidelines define agency protocols for generating and posting content online. Social media posts are spontaneous by nature and too much oversight can restrict the ability of authorized employees to respond to events and address comments in a timely fashion. Because social media content is designed to be shared—and errant posts can quickly go viral—public agencies have taken steps to minimize the risks of posting inappropriate information in a very public forum.

For example, the Federal Highway Administration (2011) maintains a centralized system for content management and requires all posts to be approved by the Office of Public Affairs. Other government agencies rely on the judgment of their employees to make sure that posts conform to their policies. King County (2010), for example, simply advises employees to use discretion when posting on behalf of the agency:

> Once these comments or posts are made, they can be seen by anyone and usually cannot be deleted. Consequently, communication should include no form of profanity, obscenity, or copyright violations. Likewise, confidential or non-public information should not be shared. Employees should always consider whether it is appropriate to post an opinion, commit oneself or one's agency to a course of action, or discuss areas outside of one's expertise. If there is any question or hesitation regarding the content of a potential comment or post, it is better not to post.

The State of Utah (2009) offers similar advice: "What is published is widely accessible, not easily retractable, and will be around for a long time, so consider the content carefully."

Timing Is Everything

Transportation agencies use social media to communicate on a wide range of topics, but most posts fall into one of three broad categories:

- Information – Updates about agency services and real-time conditions. These posts may be time sensitive and may target specific markets such as riders on a certain commuter rail line or air passengers in a specific terminal.
- Engagement – Two-way conversations with riders about overall agency services and in-the-moment issues and concerns.
- Marketing – Information that educates customers and stakeholders about agency news and accomplishments. Marketing updates tend to be casual and light-hearted in tone.

Over the course of the day, agencies may post updates in any or all of these categories, but researchers at New York University (Kaufman 2012b) recommend the following ideal messaging balance based on time of day:

Rush hour: 65% service information, 30% engagement, 5% marketing
Off-peak: 40% service information, 30% engagement, 30% marketing

In these recommendations that were originally developed for transit agencies that experience clearly defined peaks, service information is concentrated during the peak commuting periods when real-time information is critical. The level of social media engagement via one-on-one conversations with followers remains constant throughout the day.

Resource Requirements

Agencies that test the social space soon discover that readers expect fresh content and personal answers to their questions. And they're in a hurry. In the consumer world, research shows that most customers expect responses from a brand within a day of posting a comment and half of Twitter users want answers within 2 hours (NM Incite 2012).

While stakeholders may have different expectations for public agencies, the real-time nature of social media led many agencies with limited staff and budget constraints to wonder how they can respond in a timely manner. In a survey of public transportation agencies conducted for the Transit Cooperative Research Program (TCRP), staffing availability was the most commonly cited barrier to using social media (Bregman 2012).

Time Commitment
The TCRP survey further asked transit agencies how many hours per month they devoted to social media. Just more than half of the agencies serving major metropolitan areas allocated at least 40 hours to social media (the equivalent of one person week per month), and 23% invested more than 80 hours per month. For agencies in small cities or rural areas, most (86%) reported monthly commitments of 40 hours or less.

Researchers at the Fels Institute of Government asked representatives from U.S. cities how much time they spent on social media per week; the research incorporated a survey and follow-up interviews (Hansen-Flaschen and Parker 2012). Consistent with the findings for transit agencies, surveyed municipalities reported spending 2 to 10 hours per week on social media activities. Most of the cities interviewed for the research (90%) monitored their social media accounts at least daily, and 65% did so on a real-time or hourly basis. It should come as no surprise then that 81% of these interview subjects said that the time commitment required for social media increased since 2009.

Staff Roles and Responsibilities
Since 2010, AASHTO has conducted annual surveys of social media usage by state departments of transportation. Its most recent survey (2012) included questions about staff resources expended on communications

activities and social media operations. Participants represented DOTs in 41 states and the District of Columbia; the size of their communications staffs ranged from 1 employee (New Hampshire and South Dakota) to 50 (Texas and Arizona).

Most said that staff roles recently changed to incorporate social media activities. Thirty state DOTs reported shifts or redistributions of responsibilities in the past year due to social media activities. Several state DOTs said their existing staff were assigned the added duty of managing one or more social media tools. The AASHTO survey also shows that state DOTs have begun to hire social media managers or at least identify social media specialists within their operations. Some 26% reported at least one staff member dedicated to social media, 17% had dedicated bloggers, and 36% had someone responsible for videos.

Not every agency has the opportunity to hire a dedicated social media manager, and many adopted strategies to stay social despite budget constraints. To help manage the time requirements, some organizations limit the number of hours during which staff members monitor or respond to social activity. For example, the Port Authority of Allegheny County which provides transit service in the Pittsburgh metropolitan area monitors its Twitter account (@pghtransit) on weekdays only, from 8 a.m. to 4:30 p.m. (and until 4:00 p.m. on Fridays). Passengers are asked to call the agency's customer service phone number outside those hours.

The Indiana Department of Transportation posts a disclaimer on its website that reads, in part, "We understand that social media is a 24/7 medium; however, our monitoring capabilities are not" (INDOT, n.d.). Despite providing service around the clock, New York City Transit does not monitor its social media activities on a 24/7 basis and posts a disclaimer on its Facebook page, "IMPORTANT NOTE: This site is not monitored on a 24/7 basis. To report an emergency, please dial 911."

Technology can also help agencies stretch their resources. Third party applications like TweetDeck and HootSuite enable users to create online dashboards to manage all their social accounts from a single location. These applications also let users schedule posts in advance as a convenient option for sharing information that is not time-sensitive. Another option for agencies is to involve college students or interns in their social media activities. Researchers at the Fels Institute of Government found that many municipalities maintained centralized control over their social media activities, but delegated basic tasks like adapting press releases to interns (Kingsley 2009).

Account Management

Social media policies often assign the responsibility for account oversight to a particular individual or department. This helps ensure that the agency speaks with a single voice across multiple social media channels and also facilitates accountability. For many public agencies, this responsibility rests with the marketing and communications department, although in some cases the chief information officer plays a role. For example, at the state level, the Minnesota Department of Transportation requires employees to obtain permission from the Office of Communications before establishing an official DOT presence on a social media site (MNDOT 2011).

Coming Full Circle

Social media can provide transportation agencies with a wealth of information about their services, networks, systems, and employees. Twitter posts can help an agency identify a chronically congested corridor. YouTube periodically features videos showing transportation workers sleeping on the job or texting behind the wheel.

An agency's first instinct may be to ignore such posts—especially when they cast the agency in a negative light—but social media can help an organization identify trouble spots. The challenge is to ensure that the information reaches people in the organization who can act on it. Kaufman (2012b) writes:

> This information cannot stop with the public affairs officers who manage the social media account; it should be quantified, evaluated, and passed on to appropriate organization members. Frequent complaints can pinpoint a reckless train conductor, an especially friendly bus driver, or bring to light an incident that has not yet been reported by management. Using incoming feedback for improvement is a major step in social media success.

Bay Area Rapid Transit (BART) in San Francisco, for example, summarizes social media posts and distributes them to staff each week. Some organizations like the Portland International Airport (@flydpx), in Oregon and TransLink (@translink) in Vancouver integrated Twitter into their customer service centers so that passenger comments submitted via social media are routed to the proper departments.

Whose Life Is It Anyway?

The growth in social media coupled with advances in mobile technology blur the line between personal and professional activities. This trend created new challenges for public agencies as they try to ensure that personal social media activities do not compromise them or their employees. To help understand these challenges, researchers at the Center for Technology in Government defined three types of social media activities in the workplace (Hrdinová et al. 2010):

- Official agency use — Employees use social media for communicating an agency's interests. Their tasks may include writing a blog update about a program or using Twitter to provide transit service updates.
- Professional use — Workers use social media to enhance their job responsibilities, perhaps by sharing information with peers or colleagues at a site like LinkedIn or researching a topic on a YouTube channel or Facebook page. Professional uses typically access external social media sites, not those maintained by the worker's employer.
- Personal use — Employees use social media for non-work activities, such as sending a tweet or updating their Facebook status.

Research confirms the growing use of social media in the workplace among public sector employees, especially for professional purposes. A 2011 survey of government workers and contractors found widespread use of social media on the job and at home (Market Connections 2011). While nearly all surveyed workers used social media at home (92%), most also accessed the platforms at work (74%) and via mobile devices (70%).

Agencies are concerned about the implications of this trend. An obvious issue is lost productivity when workers conduct personal activities during work hours. Government agencies also worry that employees may inadvertently reveal sensitive information through social media channels or create personnel issues by complaining about supervisors or subordinates online. In a brief on social media policies, the National League of Cities writes about the potential risks associated with use of personal social media accounts outside the workplace (NLC-RISC 2011):

> Employees sometimes use their personal social media to discuss their jobs and post work-related photographs or information that expose the government to liability or compromise its confidential information. Many interact with co-workers, even with their supervisors/subordinates, and

real or perceived slights, harassment, retaliation or discrimination can follow them into the workplace.

While government agencies have little control over use of personal social media outside the workplace, many exert tight control over social media activities on the job.

Employee Access to Social Media

Government agencies generally manage employee access to social media in one of two ways. Some define the numbers or types of employees who have permission to access social media. As discussed earlier in this section, communications employees are typically authorized to use social networking on behalf of their employers. Other agencies restrict the number or type of social media sites that employees may access, often using technology to block sites like Facebook and Twitter. Employers that allow on-the-job access to social media believe that these tools can facilitate research and encourage staff collaboration. Agencies that limit social media use are concerned about lost productivity and other negative consequences.

State level policies tend to restrict social media access. Only 26% of the state DOTs responding to the 2012 AASHTO survey said that agency employees had access to social media sites and tools; 64% said that workers did not have such access. The situation appears to be more relaxed at the federal level. A 2012 survey of federal workers found that 2 out of 5 federal managers (44%) were allowed to access social media on the job, up from 20% in 2010.

Among those using social media sites, 58% used Facebook, 46% watched YouTube, and 35% accessed LinkedIn (Market Connections 2012). Their reasons for using social media included more informed decision making, research and information gathering, communication with citizens, colleagues, and other agencies, and promotion or marketing (Market Connections 2011).

Acceptable Use

In addition to defining which workers may access social media on the job, many government agencies developed policies that define how their employees may use social media at the workplace. Washington State (2010) draws a clear line between personal and official use of social media sites. The guidance for state agencies developing social media policies includes the following language:

219

Personal versus professional use: Employees' personal social-networking sites should remain personal in nature and should not be used for work-related purposes. Employees should not use their state e-mail account or password in conjunction with a personal social networking account.

Use of state resources: Employees may not use state-owned resources (computer, network, cell phone, etc.) to access social networking websites unless authorized to do so for official use. Employees must not use any state resources to access social networking sites for political purposes, to conduct private commercial transactions, or to engage in private business activities.

Elsewhere in Washington State, King County takes a more flexible approach. The county's social media handbook which applies to workers at King County Metro Transit in Seattle allows some personal use of social media during working hours although some caveats apply (King County 2010).

> During normal business hours, employees may use personal social networking for limited family or personal communications so long as those communications do not interfere with their work and as long as they adhere to existing computer use policies. Should employees discuss their county work on personal social networking accounts or websites, they should be aware their account may be subject to public disclosure laws, even if produced on personal time and equipment.

Employee Conduct

Many agencies already have policies that address employee conduct at the workplace and several have begun to address elements specific to social media. For example, the State of Utah has very clearly stated guidelines for employee conduct online (2009). Workers are advised to identify themselves as state employees and be transparent in their online activities. To ensure that transparency does not go too far, the policy further states:

> Make sure your efforts to be transparent do not violate the State's privacy, confidentiality, and any applicable legal guidelines for external communication. Get permission to publish or report on conversations that are meant to be private or internal to the State. All statements must be true and not misleading and all claims must be substantiated and approved.

The Missouri Department of Transportation also asks workers to identify themselves when discussing MoDOT-related matters online and to represent the department in a positive way (MoDOT n.d.). Advice includes, "Respect your audience. MoDOT values diversity and other opinions. Don't use ethnic slurs, personal insults, obscenity, or engage in any conduct that would not be acceptable in MoDOT's workplace."

Institutional Considerations

Social media has implications for agencies that extend beyond the purview of the marketing and communications department. Considerations include compliance with state and local requirements for archiving and record keeping, protecting user privacy, and safeguarding the agency against viruses and malware.

Archiving and Record Keeping

Social media archiving (retention of messages posted and received on social media sites) has become a topic of concern among legal teams of agencies and local governments. Archiving concerns electronic or paper recording of communications—typically outgoing, but occasionally incoming as well—taking place on social media networks like Facebook, Twitter, wikis, and blogs.

The law is still evolving, but many organizations now treat all social media posts as matters of public information subject to the same retention rules as other official communications. Retaining social media records has several advantages for an agency. First, the records may be called upon, for example, when an agency is accused of neglecting to inform the public about a street closure or a public relations team member is suspected of posting inappropriate content.

Such requests for public records typically fall under the Freedom of Information Act (FOIA) which took effect in 1967. FOIA provides that "any person has a right, enforceable in court, to obtain access to federal agency records, except to the extent that such records (or portions of them) are protected from public disclosure by one of nine exemptions or by one of three special law enforcement record exclusions" (USDOJ 2011). Whether social media posts such as tweets, Facebook updates, Foursquare check-ins, and their accompanying retweets, likes, and comments are public records subject to FOIA is still up for debate. Experts often recommend treating social media posts as public records because they are just a step away from e-mails which are already considered public records (FCC 2013).

Social media records allow agencies to look back and measure their performance in public information areas by asking questions like "Did we post service notices frequently enough?" and "How many people clicked on our links during various times of day?" Evaluating their performance on social media will assist agencies in providing more informative and engaging content.

Records also track incidents like account hacking. If an agency's security has been breached and a hacker posts messages using the account name, the agency will have records of the breach, both for proof of misconduct and to pinpoint the moment of security lapse.

To summarize the objectives, the American Council for Technology and the Industry Advisory Council (ACT-IAC) describe social media records retention as "evidence of the organization, functions, policies, decisions, procedures, operations, or other activities of the Government or because of the informational value of data in them" (2011).

Challenges

Several challenges specific to social media archiving have arisen, often due to varied interpretations of "records" and perceived needs to access archives in the future. Some of those challenges were expressed in the ACT-IAC study titled "Best Practices Study of Social Media Records Policies" (2011):

- Do "records" constitute both outgoing and incoming messages? That is, must reader comments on an agency's blog be retained?
- What needs recording? If a message exists in the public domain (such as on Twitter), must it be retained by an agency or is Twitter's existing record sufficient?
- How can metadata—tags that identify content after storage for subsequent retrieval—be applied to social media content automatically? Currently, agencies tend to assign metadata manually, which is cumbersome.
- Does the agency provide the necessary staff training and resources to conduct social media archiving?
- Will these archiving rules be enforced, and how?

Additional challenges cited by the National Archives and Records Administration (NARA), (2010) include:

- Public expectations that archived content is readily available
- Content must be drawn from multiple sites and streams

- Ownership of posts on third party sites
- Retaining interactive features that cannot be printed
- Retention schedules including timelines for deletion
- Frequency of recording
- Management of records containing personally identifiable information

Policies

Although the challenges listed above are significant, no national policy exists for social media records retention (except requirements for federal agencies). NARA developed several guidance documents, and many state and local agencies created their own policies. The resulting regulations vary widely, but typically aim to address the questions discussed below.

Content: What forms of communications must be recorded? Do they include both agency- and user-created content?

NARA (2010) defines social media content as a federal record if one of the following is fulfilled:

- Is the information unique and not available anywhere else?
- Does it contain evidence of an agency's policies, business, mission, etc.?
- Is this tool used in relation to the agency's work?
- Is use of the tool authorized by the agency?
- Is there a business need for the information?

Under these guidelines, social media posts are typically viable records, and archives should include both agency- and user-generated content. Agencies often notify users that their social media activities are considered public records. The Washington Metropolitan Area Transit Authority (WMATA) explicitly states in its Social Media Terms of Use: "All communication via our social media channels is considered public. Posts as well as any feedback from the general public will become part of the public record and will be subject to applicable retention."

The Massachusetts Department of Transportation takes the public disclaimer further by noting on its website that communications will be recorded and may be subsequently disseminated:

> Please note that the Records Retention Law of the Commonwealth requires MassDOT to preserve records created or received by a state employee. Pursuant to this retention requirement comments posted or messages received via an official state agency page on a third party

website (such as an official agency profile on a social network) will be treated as state governmental records and may be permanently archived. Information that you submit voluntarily through social media sites associated with this agency where such information is publicly available, including your name, city or town, and the substance of anything that you post may be disseminated further by being posted online at this website or be publicly discussed by a member of the administration.

Methods: What methods are used to record social media content?

Agencies often use a variety of methods, ranging from paper-based to automatic recording. The City of Seattle, for example, automatically records outgoing tweets in a non-public blog on the shared CityLink system and records the list of followers on a quarterly basis (2009b). Facebook posts are entrusted to the site's application programming interface (API), but must be printed out if they cannot be recalled using this tool (2009a).

The City of New York opted to archive all its social media content in the Nextpoint third party cloud preservation system that automatically archives more than 100 social media feeds from 40 city agencies. The service maintains a searchable archive accessible by city employees (Waldman 2011). However, unlike WMATA and MassDOT, New York City does not explicitly state in its customer use policy that content will be archived (City of New York 2011).

Several other third party systems are available for archiving and are discussed later in this section. However, it should be noted that some agencies reported anecdotally that they also retain paper copies of social media content, for fear of a lack of access to the cloud or folding of the cloud hosting company.

Access: How will archives be accessed?

While some governments like New York City provide archive access exclusively to employees, North Carolina recently drew attention for providing all social media archives to the public on a searchable website. As of December 2012, nc.gov.archivesocial.com contained 55,000 records of Facebook and Twitter communications from select state agencies as part of a pilot program. North Carolina created the site with ArchiveSocial (an archiving company) with the intent to "capture the full context of social media as it happens, and make the records almost instantly available to the general public." At the time of writing, North Carolina's system appears to be the first and only of its kind (PRWeb 2012).

When agencies allow archive access exclusively to employees, they still require FOIA requests to release the data to third parties. The employee expenditures of time and effort to search for and retrieve specific records may lead more government agencies to opt for publicly searchable solutions like that implemented in North Carolina.

Shelf Life: How long must records be retained?

The retaining of social media records must adhere to state or local policies that typically specify periods for which archives must be made available. For example, Washington State ensures retention of electronic copies of all inbound and outbound social media communications. Within the state, records may be destroyed after 6 years or fewer if 6 years is seen as "both unnecessary and uneconomical" (Washington State Legislature n.d.).

Enforcement: Will these archiving standards be enforced?

Policies do not typically address enforcement of archiving rules. An industry survey found that many organizations do not take archiving seriously: "38% of those polled admit that there is little or no enforcement of their records management policies and 55% set no guidance on dealing with important e-mails as records" (Skjekkeland 2009). When developing archiving requirements, it is essential that policymakers consider the ramifications, if any, of an absence of required archives.

Archiving Tools

Several third party archiving tools were recommended by Washington State and others and are presented in this non-exhaustive list:

- ArchiveSocial, Nextpoint, and Smarsh preserve social media feeds and provide cloud-based searchable records.
- TwInbox integrates Twitter with Microsoft Outlook, providing automated Twitter records.
- ArchiveFacebook saves Facebook content to a user's hard drive.
- SocialSafe saves records of social media streams from various networks onto a user's hard drive.
- HootSuite is social media management software that archives activity at different levels based on membership type.

Cyber Security

Social networking sites can expose public agencies to cyber threats and malicious content. The New York State Office of Cyber Security and Critical Infrastructure Coordination identifies some of these risks (Pelgrin 2010):

> Many social media sites do not have adequate security controls to pro-tect the information they are holding. For example, some sites do not require strong passwords, some transmit credentials in clear text, and some use easily guessed "secret" or "challenge" questions. As a result, social media accounts are frequently compromised. If the same account credentials are used for both the external social media site and State resources, this could lead to unauthorized access to State resources.

To mitigate these threats, the state issued guidelines for secure use of social media, focusing on governance, technological controls, and policy controls. For example, New York State agencies are required to develop content management processes as follows:

> A process must be established for reviewing and approving updates to publicly available content. These reviews must consider the type of information being made available, the accuracy of the information, and potential legal implications of providing the information, such as confi-dentiality and copyright issues.

The State of North Carolina also addresses security in its social media guidelines (2009). The state first acknowledges that many threats result from user behavior, not technology, and that employee training can help minimize the risks:

> It is important to note that security related to social media is fundamen-tally a behavioral, not a technology issue. In general, employees unwit-tingly providing information to third parties pose a risk to the core state network. Consequently, employees should know the major threats they may face and how to avoid falling prey. Prevalent social media security risks include third party spear phishing, social engineering, spoofing, and web applet attacks.

To minimize the amount of information a hacker can gain from a single attack, employees are urged not to duplicate their user IDs and passwords across multiple sites, and agencies are encouraged to provide security awareness and training to their online staffers.

User Privacy

Government use of social media has raised concerns about protecting user privacy. Public agencies frequently have procedures in place that address the use of personal information on their own websites but the same protections do not apply to channels like Facebook or Twitter. Such third party platforms are usually governed by their own privacy policies.

When public agencies use social media to disseminate information, the impacts on user privacy are minimal. However, organizations should exercise great care when interacting with citizens through social media platforms. When individuals post comments or share content on a social media site, they may also be sharing personal information and in some cases, information about their friends and contacts with an agency and its online community. Individuals often have the option to modify the privacy settings on a specific application, but not all users take the time to do so. Accordingly, social media experts encourage public agencies to minimize the amount of personal information they collect via social media and remind users that their information may be publicly visible.

Metro Transit of St. Louis posts a privacy policy on its NextStop STL blog that outlines how the agency handles personally identifiable information (2012). Here is how King County's social media policy addresses such concerns:

> Communication through agency-related social media is considered a public record and will be managed as such. All comments or posts made to County agency account walls or pages are public, not private. This means that both the posts of the employee administrator and any feedback by other employees or non-employees, including citizens, will become part of the public record.
>
> Because citizens may not realize that their social media postings are part of the public record, county agencies are encouraged to post a disclaimer to remind users their posts may be subject to monitoring and disclosure.

Conclusions

In addition to its many advantages, social media brings challenges, and some transportation agencies have been wary about adopting these new communications tools. Agencies have raised questions about getting started with social media, staffing and resource requirements, managing employee use of social media, and the legal and institutional ramifications.

Agencies increasingly rely on social media policies and protocols to manage their activities and provide guidance to employees. Some organizations are covered by existing federal, state, or local policies; others developed guidelines specific to their own agencies. Typically the policies

227

define which employees may access social media accounts on behalf of the agency, how they may use the platforms, and what they can and cannot say online. Many agencies also have policies in place that define record keeping and archiving rules for social media posts and outline steps to protect user privacy and agency cyber security.

RULES OF ENGAGEMENT: USING SOCIAL MEDIA TO TURN CRITICS INTO FANS
Susan Bregman

Social media allows transportation providers to build relationships with citizens and stakeholders. Through one-on-one conversations and interactive content, transportation agencies are learning to engage their constituents on a personal level. When things are going well—when traffic is normal, airplanes are taking off, and trains are running on time—agencies may see a lot of shares, retweets, and "likes." But when things go wrong—a train derails or a tornado shuts down an airport—the tone of the conversation can quickly change from happy talk to profanity laced diatribes. This section talks about using social media to engage with the community and ways to respond when things (inevitably) go wrong.

What Is Engagement?

Social media experts talk a lot about engagement, but can't agree on a definition. Pundits dismiss the term as just another buzzword in an industry that already generates a lot of buzz. Some see engagement as the process of connecting with an audience, while others treat engagement as an end goal, something to be measured by counting clicks, shares, and retweets.

Writing in the *Social Media Explorer* blog, industry expert Jason Falls describes engagement as the process of building relationships with customers. Although his focus is on selling products, not providing services, the core definition applies to public agencies as well. "Engagement is not a goal. It's a result," he writes. "When you join the conversation, make your communications focused on the consumers and not your brand; when you build relationships, not billings, engagement happens" (Falls 2012).

In his book titled *Engage!* Brian Solis talks about the importance of engagement in building relationships through social media interaction. He writes, "Active listening and the resulting informed engagement plant

228

the seeds that flourish into meaningful conversations and relationships" (Solis 2010).

For agencies, engagement comes from building relationships with members of their social media circles, often one citizen at a time. It is no longer enough for agencies to use the social space to distribute press releases and post an occasional instructional video on YouTube. Customers and stakeholders expect more, as this advice for public educators suggests (Reform Support Network 2012):

> Audiences on social media expect that if an agency is using Facebook or Twitter, it is ready to engage with members of the public, not just speak to them. Although posting materials, statements, press releases and newsletters is important and will help to educate key audiences, these efforts are not typically enough to get audiences to engage in a conversation or offer feedback about key topics.

In the transportation world, engagement ultimately leads to better informed customers and better service (Kaufman 2012b):

> Customer engagement takes many forms: reposting news from others, responding to customer needs, asking questions, and appealing to the broader group. Engagement leads to more "likes," "fans," "followers," and "subscribers"—the numbers of which are important for perpetuating important messaging. When more travelers are informed about potential service changes, they will be motivated to change their plans as needed and amplify the message to other affected travelers, overall ensuring that as many travelers as possible are moving efficiently and safely through the system.

Road to Engagement

The agencies that use social media most successfully actively seek out opportunities to connect with people in the social space. Industry experts agree that two elements are key to building relationships with followers. First, post useful and interesting information, with an emphasis on multimedia content like photographs, videos, and infographics. Second, engage members of the community by asking questions and listening to their responses.

In a brief on best practices in government use of social media, *GovDelivery* (n.d.) offers advice to practitioners: "Beyond posting multimedia, recent studies have shown that asking questions and engaging in a two-way dialogue not only increases your reach but deepens your relationship with your followers and fans." Photos and videos can "help illustrate concepts like construction work, explain policies through discussion, and demonstrate use of the system," all of which in turn can engage and inform riders. And asking questions of the audience can both provide information to the agency (about a broken ticket vending machine, for example) and keep customers engaged.

Not everyone heeds this advice. An assessment of Twitter use among transportation providers in the New York metropolitan area found that agencies were neither posting dynamic content nor engaging customers by asking questions. "By not asking questions, transportation providers are missing opportunities to learn from their audiences where information is needed, and to foster positive relationships" (Kaufman 2012a).

How can agencies make the transition from one-way posts to two-way communication? In guidance prepared for civil servants, the British government defined the social media engagement cycle (Government Digital Service 2012). The path to social media engagement can start very narrowly, as individuals simply follow online conversations in a particular area of interest. However, as Figure 6.3 shows, the path eventually broadens and can ultimately lead users into conversations about a wide range of topics that extend far beyond the initial focus of concern.

The Washington Metropolitan Area Transit Authority is a textbook example of an agency that traveled the path from static communications to active engagement. Metro initially used its Twitter account as a

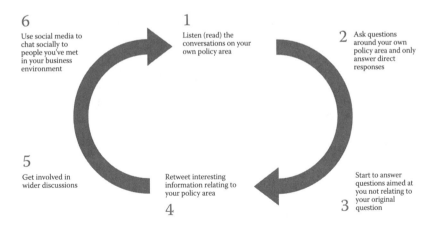

Figure 6.3 The path to social media engagement starts with listening. (From: Government Digital Service, Cabinet Office. 2012. Social Media Guidance for Civil Servants. Part 1: Guidance on the Use of Social Media.)

one-way street, pushing out automated service alerts almost exclusively. This changed in 2011 when the agency hired a new spokesperson who redefined the agency's approach to active engagement. Suddenly Metro started having conversations with riders. Spokesperson Dan Stessel started by talking about Metro Forward, a major capital improvement program, but in some ways the topic of the conversation was irrelevant. A key goal of Metro's new approach was to build trust with its riders. A profile of Stessel in *Washington City Paper* describes the change (Desantis 2011):

> By engaging with riders through social media channels, Metro won't be able to shape the message all the time, but Stessel isn't interested in begging riders to believe what he's telling them. "I'm here to demonstrate that we're worthy of being believed," he says.

Engagement with customers is now part of Metro's culture, embedded in its online social media policy (WMATA n.d.) which reads, in part, "WMATA's social media channels are an opportunity for us to engage in dialogue with our customers and the general public. These channels are managed by WMATA employees and are not automated feeds." Is it working? When Dan Stessel came on board, Metro's Twitter account, then called @MetroOpensDoors, had about 7,000 followers. In June 2013, more than 46,000 people follow Metro's Twitter account (now @wmata).

Turning Frowns Upside Down

Above all, social media is interactive. While organizations can control the information they send out, they cannot control what people say about them. Although Facebook has yet to create an "unlike" button—perhaps for good reason—social media activity is not all sunshine and lollipops.

Transportation agencies are no strangers to criticism—online and face-to-face. Transit riders complain about late buses and crowded trains. Airport patrons fume about long check-in lines. Neighborhood residents protest a planned highway ramp. Individuals increasingly turn to social media to express their anger. The immediacy and anonymity of social media can invite comments that are unfiltered and visceral. Unfortunately, they are also very public. This fear of online criticism was one of the top barriers to using social media among transit agencies surveyed in 2011 (Bregman 2012).

This fear seems to be largely unfounded. Following the leads of private sector companies, some agencies discovered that reaching out to critics can be the best way to neutralize online negativity. As officials at the Washington State Department of Transportation and Arizona Department of Transportation recounted in Chapter 3, sometimes a direct response to a negative comment can defuse a tense situation and occasionally transform an online critic into a supporter.

A Tale of Two Systems

When Hurricane Sandy created havoc along the East Coast in 2012, transportation agencies used social media to keep their customers informed of the changing and challenging conditions. Some agencies rose to the occasion and provided the information their stakeholders needed, while others left citizens frustrated and in the dark. In a post-Sandy assessment, *The New York Times* compared the social media efforts of the Long Island Rail Road (LIRR) and the New Jersey Transit Corporation (NJT) and discovered two very different strategies (Rivera 2012).

LIRR posted photographs and videos on its Facebook page before, during, and after the storm. Readers could see the devastation firsthand as LIRR shared images of everything from downed trees to an unmoored boat that landed at a rail station.* The agency provided service updates, answered reader questions, and posted links to external resources. Videos documented the washed-out sections of the rail bed and presented interviews with grateful customers as soon as partial service was restored.

NJT also updated its Facebook page to keep riders apprised of service changes but, according to *The New York Times*, the agency did not start posting photographs until the storm was gone. Perhaps more troubling, staffers answered rider questions intermittently and sometimes incorrectly.† *The New York Times* described the result:

> The Long Island Rail Road conveyed a narrative of shared pain, of workers fighting back against unprecedented damage that was beyond their control. Passengers frequently and vociferously critical of the railroad suddenly sympathized and even praised communication efforts that, if not perfect, were viewed as improved. New Jersey Transit Corporation communications, on the other hand, became for many commuters yet another source of misery.

The New York Times also took an unscientific look at the level of customer engagement on each agency's Facebook page during a nine-day

---

* https://www.facebook.com/mtalirr
† https://www.facebook.com/NJTRANSIT

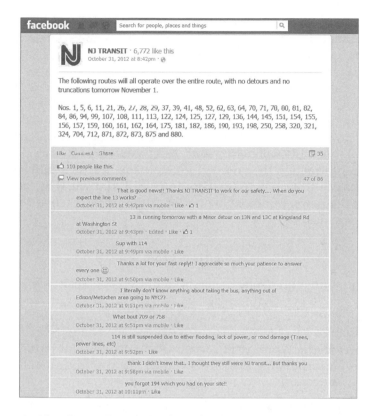

Figure 6.4 New Jersey Transit used Facebook to provide service updates after Hurricane Sandy. (Screen capture from Facebook page https://www.facebook.com/NJTransit).

period. Using an unofficial "confusion quotient," calculated as the ratio of agency answers to customer questions, the newspaper compared levels of customer engagement of the two agencies. LIRR posted an answer for every two questions or rider posts that shared information, a ratio of approximately 50%. NJT, in contrast, responded to only 13% of rider queries or related posts—just over one answer for every eight questions.

Underscoring the limited updates available from NJT, a former employee jumped in to fill the apparent vacuum left by the agency and took it upon himself to dispense service updates to information-hungry passengers. Figure 6.4 shows a series of service updates on NJT's Facebook page, including comments from the former worker.

Both organizations received similar shares of negative comments on their respective Facebook pages, but the tone and content differed. LIRR riders focused on service issues like overcrowding as service slowly returned, while NJT riders were frustrated about the lack of accurate schedule information for modified bus and train services. Both agencies suffered significant damage from Hurricane Sandy, but each agency made different decisions about its response to the storm in the social space. Neither NJT nor LIRR offered perfect service in the aftermath of the storm, but LIRR connected with its customers and equipped them with the information they needed to make the best of a difficult situation. Ultimately, keeping customers informed and engaged, no matter how bad the news, can be as important as the transportation service an agency provides.

Navigating Mishaps

Sometimes things go wrong in the social space. An employee might inadvertently post a personal message on an agency Twitter account. A joke might fall flat or worse, offend someone. Citizens may launch an online protest against an agency policy or decision. The New Zealand government developed a typology of social media faux pas and recommended strategies for responding to them. Social media missteps can fall into the following categories:

- Directed abuse: Abuse or criticism directed at the agency or a worker
- Leak or early release: Information distributed before its official release date
- Unwanted intrusion: Unwelcome participation in a social media conversation
- Heated topic: Discussion of a controversial or partisan topic
- Misinterpreted message: Incorrectly interpreted message
- Misaligned expectations: The audience and agency have different expectations
- Hack or wrong account posting: Someone posts inappropriate messages on an agency account or a user accidentally posts personal messages on the agency account
- Questionable humor: Attempted humor that the audience does not find funny
- Insensitive statement or opinion: A post that the audience perceives as insensitive or offensive

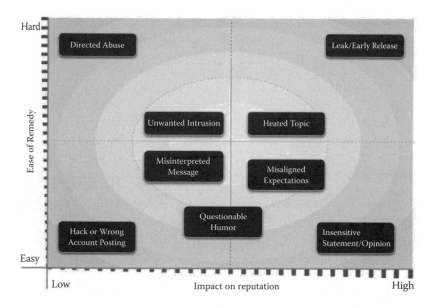

Figure 6.5 Social media mishaps can range from questionable humor to directed abuse from outside parties. (From: New Zealand Department of Internal Affairs. 2012. Social Media in Government: How to Handle a Mishap, Version 1.0, p. 4.)

Some of these missteps are more damaging than others, and some can be especially difficult to fix. To help agencies better understand the impacts and risks, the New Zealand government (2012) rated these mishaps based on two criteria (1) the extent of impact on brand and reputation, and (2) how easily the result can be fixed. Figure 6.5 presents this typology. Protocols are in place for addressing some of these issues which range from a hack or wrong account posting (low impact, easily remedied) to leaked information (high impact, difficult to remedy). Some examples are presented here.

Hack or wrong account posting (low impact/easy to fix) — If an agency account is hacked, the agency should contact the social media provider as soon as possible and review internal login and authentication procedures. Accidental postings require a different response. Social media tools like TweetDeck and HootSuite allow users to manage multiple social media accounts from a single dashboard; they also make it easy to mix up accounts. If a worker inadvertently posts a personal update on an agency page, a brief apology is appropriate. Figure 6.6 shows an example

Figure 6.6 TBARTA apologizes for double-tweeting an update. Such errors have little impact on an agency and are easy to fix with a simple apology. (Screen capture from @TBARTA.)

of an informal apology from the Tampa Bay Area Regional Transportation Authority after an errant tweet.

Directed abuse (low impact/hard to fix) — Online abuse directed at an agency or an individual employee can be difficult to address. In extreme cases, the posts may violate the terms of service of the platform and the service provider can investigate the situation. At other times, contacting critical individuals en masse or on a one-to-one basis can help defuse a situation. Agencies under attack can sometimes change the tone of a conversation by reminding angry posters that real people are behind the social media account. Sometimes, especially when an agency has worked to develop good relationships with its followers, other members of the online community will step in to defend the organization against abusers.

In 2011, Calgary Transit had a bad day and used Twitter to alert riders about service delays. The service disruptions lasted for hours and Calgarians used Twitter to share their frustration. After hours of abuse, Calgary Transit posted the following message on its Twitter account (@calgarytransit): "When you tweet please consider we are humans at this end. Our staff will not tolerate abuse under any circumstances. Thanks." The agency followed this tweet with an expression of gratitude: "We appreciate everyone's understanding, patience, and fairness. Thank you! #yyc #yyctransit." This helped turn the situation around and citizens responded with messages of support. Even Calgary's mayor stepped into the fray to defend the agency. He tweeted: "I know it's been a rough couple of days on @calgarytransit. I also know they are working hard to fix it. Feedback = good! Abuse not called for" (ctvcalgary.ca 2011).

Insensitive opinion or statement (high impact/easy to fix) — As agency employees become comfortable engaging with followers in the social space, it becomes easy to make an ill-timed remark or to post a snarky comment that causes offense. In these cases, agencies should apologize quickly and sincerely.

Leak or early release (high impact/hard to fix) — This is the most challenging type of incident for agencies to address. Inadvertently releasing confidential information can compromise operations. In 2010, the British military was forced to cancel secret operations after Ministry of Defense employees leaked sensitive information via Facebook and Twitter. To help prevent such incidents, the U.S. Army's social media policy instructs soldiers to disable the GPS function on their smartphones to avoid geotagging social media posts from the field (Einhorn 2012). Posts that release sensitive information will likely shine a spotlight on an agency and officials should be prepared to respond to questions about the incident authoritatively and consistently.

Other types of social media mishaps will likely be less extreme and the appropriate response will depend on the specifics of the incident and the agency. No matter what the issue, government workers are advised first to inform agency officials and then act quickly, be honest, and apologize for mistakes. After the initial damage control, public servants can shift their attention to staying in touch with constituents, individually or in general, and keeping them apprised of agency actions to redress the incident. Throughout any incident, the recommended approach requires civility, openness, and honesty.

The U.S. Air Force also developed guidance that can help agencies navigate the sometimes tricky terrain of social media (2009). Using a flow chart format to help decision makers, the USAF response assessment has applications for all public agencies facing online criticism (Figure 6.7). The response assessment has three elements: (1) discover, (2) evaluate, and (3) respond. Within each element, an agency can choose a response based on the information at hand.

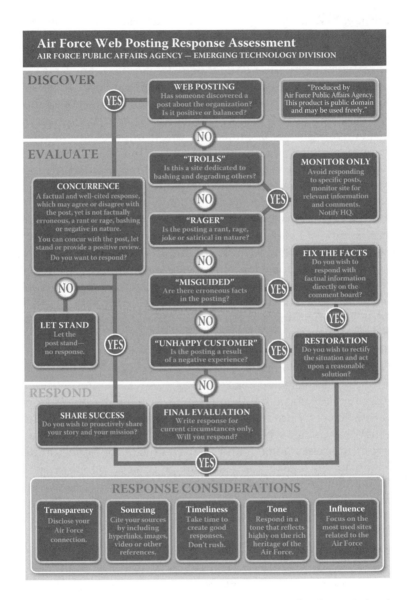

Figure 6.7 The U.S. Air Force developed an approach for determining how to respond to online comments. (From: Air Force Public Affairs Agency.)

Discover
The discovery phase of the process is simple and straightforward. After someone finds a post about the organization, the organization must determine whether it is positive and balanced.

Evaluate
If the post is positive, an agency may choose to concur with the post, let it stand, or share the information. For negative posts, the organization should first determine the nature of the criticism. Does it come from a troll (website or individual dedicated to bashing others)? Is it a rant or a joke? In both cases the USAF recommends keeping an eye on the site while ignoring the specific post. Is the post misguided or factually incorrect? If so, an agency may want to correct the record. Is the post from an unhappy customer? This type of post gives an agency the opportunity to make things right.

Respond
After assessing a situation, an agency can choose an appropriate response to the negative post. Considerations include timeliness, transparency, and tone. An agency may further choose to focus its attentions on influential websites and social media platforms and provide links to any relevant sources.

Conclusions

Social media allows transportation providers to engage their constituents on a personal level. These one-on-one connections can help personalize an otherwise faceless bureaucracy and enable agencies to build community support for key programs and initiatives.

The true test of an agency's social media presence is when things go wrong. An Internet troll will provoke controversy, a misinformed stakeholder will post incorrect information, or a disgruntled citizen will air a complaint in the social space. While the temptation may be to ignore such comments, savvy agencies can assess a situation and respond accordingly. In many cases, a one-on-one conversation can help resolve the issue. This level of personal engagement allows commenters to see the agency as a group of people doing their (sometimes fallible) best, rather than a pre-programmed machine that is unresponsive to the public's needs. This shift in perspective can often neutralize a critic and occasionally create a new supporter.

BRIDGING THE DIGITAL DIVIDE: ENSURING INFORMATION EQUITY IN SOCIAL MEDIA

Kari Edison Watkins, Katharine Hunter-Zaworski, and Sarah Windmiller

Whether the intent is to advocate for new service, send an alert about a current disruption, or obtain feedback about a long-range plan, transportation agencies need to reach current and potential users of the system. Although it is impossible to reach all constituents at all times (even through social media), there must be a concerted effort on the part of agencies to reach a representative audience with regard to gender, age, disability, race, ethnicity, income, and education. If outreach or feedback is biased away from one group, the resulting plans and services may not provide equal access to future transportation resources.

Why Equity Matters

Just as social and economic backgrounds are not equally distributed among transportation system users, any benefits or hindrances resulting from agency decisions or policies are not equally distributed among the population. These decisions may provide unequal access and resources not only for current users but for future generations. To ensure these decisions take into account the needs of particular demographic groups, a broad-based sample of the population must be consulted.

The need for feedback from a broad audience is essential because all users must have the opportunity and means to voice their opinions, needs, and desires. Windy Cooler's description of her region's metropolitan planning organization (MPO) summarizes the possible outcome of unequal representation (Bullard 2003):

> In a city that is 50% African-American where historically and even today, the black community is so egregiously underserved and largely unheard, and where citizens, regardless of color are uninvolved and uneducated in the [transportation] planning process, it is no wonder, in fact it is inevitable, that the needs of the few, who are powerful for the moment, are put above the needs of the whole.

Clearly, broad audience feedback is required to produce outcomes and solutions that benefit transportation communities as a whole, rather than concentrating on specific socioeconomic groups. According to Todd Litman (2010) in his equity guidance for transportation planning, equity impacts transportation in many ways:

- The quality of transportation available affects people's opportunities and quality of life.
- Transport planning decisions affect the location and type of development that occurs in an area, and therefore accessibility, land values, and developer profits.
- Transport facilities, activities, and services impose various indirect and external costs, such as congestion delay and accident risk imposed on other road users, infrastructure costs not funded through user fees, pollution, and undesirable land use impacts.
- Transport expenditures represent a major share of most household, business, and government expenditures. Price structures can significantly affect financial burdens.
- Transport facilities require significant amounts of land that is generally exempt from rent and taxes, representing an additional but hidden subsidy of transport activity.
- Transport planning decisions can stimulate employment and economic development which have distributional impacts.

Reaching constituents of all races, origins, and incomes is not only a noble goal. It is the law. Beginning in 1994 with President Bill Clinton's *Executive Order 12898, Federal Actions to Address Environmental Justice in Minority Populations and Low-Income Populations,* federal agencies have had a directive to address disproportionate human health and environmental effects on minority and low-income populations. The order was intended to provide improved access to public participation venues and public information for these communities. The directive works to ensure that: "(1) people have an opportunity to participate in decisions about activities that may affect their environment and/or health; (2) the public's contribution can influence the regulatory agency's decision; (3) their concerns will be considered in the decision making process; and (4) the decision makers seek out and facilitate the involvement of those potentially affected" (U.S. EPA 2012).

Lawsuits against transit agencies and government entities across the country often result from inequality arising from lack of access, disproportionate use of funds, and overall discrimination (Bullard 2003). Utilizing social media as an additional source for better representative audience feedback could perhaps also help agencies avoid various legal issues, lawsuits, and boycotts due to under-representation.

In addition, the legal basis of Title VI of the Civil Rights Act of 1964 dictates that no person in the United States shall, on any grounds, be

241

excluded from participation in, be denied benefits of, or be otherwise subjected to discrimination under any program or activity that receives federal financial assistance. This applies to the minority and low-income populations targeted in Executive Order 12898 and also to discrimination based on age, gender, national origin, and disability.

The Limited English Proficiency (LEP) clearinghouse, an interagency governmental working group, recommends that any government agencies that serve populations that do not understand or speak English should develop programs and hire employees who serve these language needs. Documents containing vital information such as applications, consent and complaint forms, notices or rights, and written tests for federal services and/or benefits must be translated. Many federal, state, and local transportation agencies, such as the Federal Transit Administration, Washington State DOT, Louisiana DOT, New Jersey DOT, and New York City DOT already produced LEP guidelines. Although most social media communications do not yet fall within LEP requirements, it is important to consider language needs because communication may one day be included in the regulations.

Following the intent of such legislation, all people should be able to access information. While most of the rules do not explicitly address social media, the spirit of existing legislation is clear. For transportation, this means ensuring that travelers can access customer information at every point on their journeys through printed materials, electronic and static station signage, and announcements in vehicles and at stations or stops. It is best for agencies to remember that every traveler is an individual with unique capabilities; and that different travelers may access information in different ways.

To help transportation agencies address these differing information requirements, the U.S. Americans with Disabilities Act (ADA) has long required the use of dual mode communication systems that present messages in both audio and visual formats. However, the implementation of this requirement has been inconsistent, and people with sensory and intellectual impairments are often left out.

The United Nations Convention on the Rights of Persons with Disabilities that became effective in May 2008 enshrines the principle that persons with all disabilities must be able to enjoy basic human rights and fundamental freedoms. For the first time, an international human rights agreement includes an explicit articulation of the rights of persons with disabilities to access information and communication technologies and systems on an equal basis with others and without discrimination (CIS 2012).

The U.S. also has a range of general and specific laws mandating telecommunication access for persons with disabilities. Through the Hearing Aid Compatibility (HAC) Act of 1988, the Federal Communications Commission (FCC) ensures that all telephones manufactured or imported for use in the U.S. and all "essential telephones" are hearing aid-compatible. The FCC also extended this requirement to wireless and mobile telephones. In October 2010, the U.S. Congress passed the Twenty-First Century Communications and Video Accessibility Act (CVAA) whose objective is to improve access to "advanced communications" and consumer-generated media for persons with disabilities. Some of the key requirements of the CVAA include:

- Access to Internet browsers on mobile phones
- Improved accountability and enforcement
- Expansion of relay services definitions and contributions
- Equipment for low income deaf–blind individuals
- Expansion of hearing aid compatibility rules
- Ensuring access to next generation 911 services

Section 508, a provision of the U.S. Rehabilitation Act, mandates that electronic and information technology funded, developed, or used by federal agencies should be accessible to people with disabilities. Although Section 508 concerns the federal government, it created a marketplace incentive for the development of accessible information and communication technology (CIS 2012).

New Digital Divide

If reaching all constituent groups equitably is a goal, the immediate question is whether social media platforms facilitate this goal. Several organizations track social media usage over time and among varying demographic groups. The Pew Internet & American Life Project of the Pew Research Center regularly conducts surveys of Americans' use of the Internet, allowing researchers to track usage of multiple Internet tools over time. Quantcast provides the estimated traffic and demographics of users and shows how the distributions compare to the average Internet distribution. Experian Marketing Services' Hitwise displays the current top social media websites. Another organization displaying demographic breakdowns of users is Google Ad Planner, which compiles age, gender, and traffic statistics. All these sources have been used to provide background on social media demographics in this section.

The concept of a digital divide was historically concerned with lower income and non-white ethnic groups having unequal access to telephone and computing power at home and at work, thus creating two cultures (NTIA 1995). The proliferation of smart phones rapidly changed this dynamic, with Internet access via a cell phone higher among blacks (57%) and Hispanics (64%) compared to whites (49%) as early as 2009 (Smith 2010). Similarly, the highest increase in smart phone use is among those making less than $30,000 per year (34% in February 2012 compared to 22% in May 2011).

Perhaps surprisingly, the Pew Research Center's comprehensive 2012 social networking site report finds no statistically significant differences in use of social networking based on race or ethnicity, household income, education level, or urban, suburban, or rural residence (Duggan and Brenner 2013). However, women slightly outnumber men with 71% of women using social networking versus 62% of men. This lack of a disparity among ethnic groups would seem to run counter to prevailing public perception. The results of the survey are shown in Table 6.1.

The biggest disparity in social media usage is in age. Figure 6.8 shows that 86% of Internet users ages 18 to 29 use social media, 72% of those from 30 to 49, 50% of those from 50 to 64, and 34% of those over 65. The distinct drop off in social media usage by age group is statistically significant. It is of note that the statistics are among "Internet users" and as shown in Figure 6.9, just over half of the 65-and-over population uses the Internet. This compares to younger groups with a nearly 100% Internet penetration rate, thus making these statistics conservative estimates at best. Clearly, there is a discrepancy in access to social media among varying ages.

However, although social media usage among the youngest groups seems to have reached a plateau, usage by older groups continues to climb over time. Figure 6.10 shows the trends in social media usage from 2005 to 2012. This trend is expected to continue as younger age groups whose use hits a plateau at 80% and above eventually age into the other categories.

Turning to specific social media platforms, Twitter use specifically was significantly higher among those 18 to 29 (27%) than older groups (30 to 49 at 16%, 50 to 64 at 10% and 65 and over at 2%) according to Duggan and Brenner (2013). Pew researchers also showed that Twitter use was more than twice as high among black non-Hispanics (28%) compared with other racial groups (12 to 14%). Other categories do not statistically differ for Twitter usage except for urban (20%) versus suburban (14%) and rural (12%).

Another noticeable demographic difference is the gender of Pinterest and Instagram users. While the distribution of males and females is fairly

Table 6.1 Percentage of U.S. Online Adults Using Social Networking Sites

Demographic Group	Use of Social Networking Sites (%)
All Internet users	67
Gender	
Men	62
Women	71
Race and Ethnicity	
White, non-Hispanic	65
Black, non-Hispanic	68
Hispanic (English- and Spanish-speaking)	72
Household Income	
Less than $30,000	72
$30,000 to $49,999	65
$50,000 to $74,999	66
$75,000+	66
Education Level	
Some high school or high school grad	66
Some college	69
College+	65
Geographic Location	
Urban	70
Suburban	67
Rural	61

Source: Duggan, M. and Brenner, J. 2013 The Demographics of Social Media Users, 2012. Pew Internet & American Life Project. With permission.

even among Facebook, LinkedIn, Tumblr, and Twitter users, Pew's 2012 survey found that approximately 80% of Pinterest users are female (Rainie et al. 2012). However, this is slowly changing. According to Google Ad Planner, the distribution is 66% female and 34% male as of February 2013 (Google 2013). This suggests a possible shift in gender distribution and, due to the site's relatively young age, reminds agencies that demographic distributions change as sites become more (or less) popular.

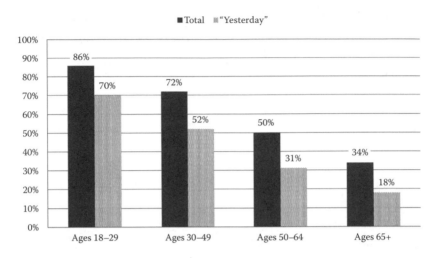

Figure 6.8 Social networking site use by age group. (Adapted from Zickuhr, K. and M. Madden. 2012. Older adults and Internet use. Pew Internet & American Life Project. With permission.)

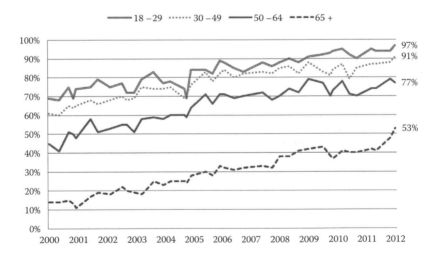

Figure 6.9 Internet use by age group, 2000–2012. (Adapted from Zickuhr, K. and M. Madden. 2012. Older adults and Internet use. Pew Internet & American Life Project. With permission.)

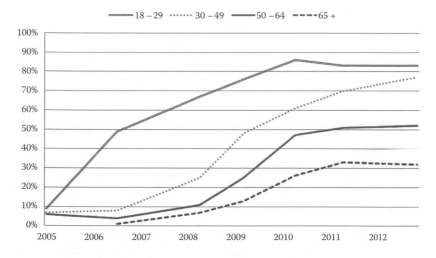

Figure 6.10 Social networking site use by age group, 2005–2012. (Adapted from Madden, M. and K. Zickuhr. 2011. Sixty-five percent of online adults use social networking sites. Pew Internet & American Life Project. With permission; and Duggan, M. and J. Brenner, J. 2013. The demographics of social media users, 2012. Pew Internet & American Life Project. With permission.)

Regarding professional social networking sites, Pew researchers found that the majority of LinkedIn users are 30 to 49 years old (38%) or 50 to 64 (34%). The percentage of LinkedIn users between 18 and 29 (15%) is the lowest among all other sites analyzed (Rainie et al. 2012). Facebook, Twitter, and Pinterest all tend toward this youngest group with usage decreasing with age (Duggan and Brenner 2013). Likewise, as income levels increase, the percentage of LinkedIn users increases as well. Approximately 9% of users earn less than $30,000 a year while 50% earn over $75,000 (Rainie et al. 2012).

Pew Research Center also found that at least 40% of the users on the four websites analyzed (Facebook, Twitter, LinkedIn, and Pinterest) had college degrees. LinkedIn ranked highest in users with college degrees (68%); the lowest percentage (40%) was found among Facebook users. Facebook also had the highest percentage (33%) of users who did not attend college, with LinkedIn ranking lowest at 10% (Rainie et al. 2012).

Despite the fact that social media usage overall is similar among all racial and ethnic groups, one major barrier remains to ensuring outreach to traditionally under-represented groups: the language barrier. In 2005, the U.S. Government Accountability Office estimated that more than 10 million people in the U.S. had limited English proficiency; either they did not speak English at all or did not speak it well. GAO further said that "these persons tend to rely on public transit more than English speakers" (GAO 2005).

People with disabilities make up another potentially underrepresented group. According to the U.S. Census (Brault 2012), 14.8% of persons aged 15 and older have severe disabilities, including 0.8% with severe seeing difficulties and 0.5% with severe hearing difficulties. Pew Research has not specifically inquired about social media usage by disabled populations. However, Pew did find that 54% of adults living with disabilities use the Internet, compared with 81% of adults without disabilities (Fox and Boyles 2012).

A research project conducted by the National Center for Accessible Media (NCAM) and the National Center for Accessible Transportation (NCAT) included a national online survey of transit industry professionals, online research, and industry outreach to review the status of passenger communication technologies, policies and practices, and the awareness and adoption of accessibility standards and specifications. The Oregon State University Survey Research Center (OSU-SRC) conducted the survey in 2012.

A key finding of the project was that the gap in accessibility was in the awareness of the need to provide accessible travel information and in the implementation and use of standards, but not in the standards themselves. The existing transit standards contain the critical elements—the capability to store and fully transform data—but data must have the right interface, and this includes presenting non-visual alternatives to items such as maps (NCAM 2012).

Social Media Benefits to Underrepresented Groups

Social media is allowing information access and public participation opportunities to many groups that were previously not reached. Making meeting materials available on an agency's website, Facebook page, or YouTube channel expands outreach to people who could not otherwise attend a public meeting because of schedule, limited transportation options, or disabilities. Many people can now read a tweet or post from a local agency. Before the use of social media, they could either not obtain such information or had to visit the agency. Social media provide traditionally under-represented groups that have Internet access the opportunities to participate in public decision-making.

Many travelers with disabilities have benefitted from using personal devices and social media platforms that present information in formats customized to their particular accessibility requirements. People who are deaf or hard of hearing were challenged during the telephone era; now they can engage with e-mail and chat services. According to Kapanci and Grant (2012), "In the social media world, a person's disability is invisible. Individuals with disabilities can engage in online communities and conversations without barriers and without judgment."

Older people and those with disabilities can now find communities of interest and advocacy groups that previously would have been difficult to engage. Social media campaigns provide disability activists with platforms to gain support and attention (Butler 2012). Finally, social networking has opened business and educational opportunities and personal connections for people with disabilities (NCD 2011; Hollier 2011).

For populations with limited English proficiency, a multitude of web-based applications can translate web pages. Examples are FreeTranslation.com, translate.google.com, bing.com/translator, and free-website-translation.com. These sites allow web developers to add translation buttons on their websites for easy access to translated pages. Facebook and many blog sites have features in the settings index that allow pages to be translated into whatever language the user chooses. Both Twitter and Facebook have online communities that help translate the main website pages (but this does not include posts or tweets).

Many transit agencies such as the Washington Metropolitan Area Transit Authority, TriMet in Portland, and Sound Transit in Seattle already have multilingual websites, while others use a Google translator to send their main pages to Google to be translated to the language of choice. This can apply to blogs, tweets, and posts as well. Agencies should use care when translating online material, however, and avoid automated translations whenever possible. The U.S. General Services Administration (GSA) guidance at HowTo.gov strongly discourages the use of machine or automatic translations even if disclaimers are used. Instead, GSA encourages agencies to have qualified language professionals review sites, posts, and tweets before posting to ensure translations are correct (US GSA 2012).

To date, most U.S. transportation agencies are not conducting social media activities in multiple languages, but a few are trending in that direction. Los Angeles Metro runs a separate Spanish language blog called *El*

Figure 6.11 A Spanish language tweet from Los Angeles Metro alerts Spanish-speaking customers about upcoming parking restrictions and bus detours. (Screen capture from @elpasajero.)

Pasajero and related Twitter account (@elpasajero); the blog shares some content with Metro's English language blog, *The Source*, but posts original materials as well. Figure 6.11 shows a tweet from *El Pasajero* alerting followers to bus detours because of the Academy Awards.

Strategies to Make Social Media More Accessible

Although the reach of social media is rapidly expanding, it is anticipated that older populations will continue to be underrepresented by social media outreach. For this reason, agency outreach needs to include strategies to overcome this bias. Fortunately, traditional public involvement measures such as meetings advertised by mail and newspaper tend to attract older populations in greater numbers. Therefore, if anything, social media is correcting a public involvement bias against younger populations. Until social media usage is nearly ubiquitous, these traditional measures of involvement must continue to be used. *Talking Transportation* (Brown 2012) writes about the role of social media in a balanced public participation plan:

> First, acknowledge that at least for now, social media tactics are likely complementary to a good core program. After all, we still have to consider those who are not online. And even with more than 750 million Facebook accounts, there still are a few people who resist the urge to read their friends' status updates and look at cute pictures of their nieces and nephews. Good social media plans rest atop solid traditional outreach programs, enhancing and enriching the opportunities for public input and, eventually, public decision-making.

With regard to specific social media platforms (Facebook versus Twitter or Pinterest versus LinkedIn), the demographic profiles of users should be taken into account when developing a plan to reach a population so that agencies are aware of any lack of uniform representation. While social networking in general and most social media platforms show little difference among income, ethnicity, and education, some platform-specific differences still exist. One example is the gender discrepancy among Pinterest users. Agencies should consider these factors when planning their social media strategies. In regions with substantial minority and LEP populations, agencies should consider making their social media posts available in appropriate languages even though they are not required to do so currently.

Several approaches are needed to ensure inclusion of all travelers in the use of social media. While agencies cannot control the applications created for specific social media sites, they can change the way information is presented. Screen readers and video captioning can improve social media functionality for some populations, but agencies must format their messages properly for these improvements to be utilized.

Social media sites that are primarily image-based are problematic for those who are blind or visually impaired. However, the technological advances within recent years have yielded one solution to this potential problem. Screen readers allow these users to have access, utilize, and contribute to these social media sites. A screen reader is software that can be installed on computers, phones, and similar devices and reads the material on device screens aloud. Some of the most popular software systems are JAWS, Window-Eyes, and Apple's VoiceOver which is pre-installed on new iPhones (WebAIM 2012; Apple 2013).

251

In May 2012, WebAIM surveyed people who used screen readers about their use of social media sites. When asked how accessible social media websites are, nearly 47% of respondents found social media sites to be "Somewhat Accessible" when using a screen reader. Approximately 25% thought the sites were "Somewhat Inaccessible" while only 7.4% and 8.7% thought the sites were "Very Accessible" or "Very Inaccessible," respectively (WebAIM 2012). In its report, WebAIM notes that the overall accessibility of social media sites is only slightly higher than accessibility in December of 2010. This indicates that more investment in improving access is needed, especially when approximately one third of respondents stated that social media sites were inaccessible. However, improving screen reader technology is not the only solution for improving a website's accessibility.

Screen readers generally convert every text element to speech, including special symbols that can create challenges in a social media environment where symbols, acronyms, shortened links, and icons are used extensively. While a tweet may seem fine initially, the message, when read by a screen reader, may become confusing and/or interpreted incorrectly. For example, if someone tweets:

> http://tinyurl.com/bgcq3ov @oaksquaresusan @transitmom are writing a #socialmedia book

it will be read as:

> HTTP colon slash slash tinyurl dot com slash BGCQ3OV at oak square susan at transit mom are writing a pound social media book

To prevent this, it is important to keep in mind what will be interpreted by a screen reader. The symbols Twitter uses, such as *at* replies (@), hashtags (#), and links, will be fully read. Each letter in a shortened link like the TinyURL link will be individually read, potentially creating an unnecessarily long and confusing tweet. Also, a voice reader may not recognize some abbreviations or acronyms and read each letter individually. For example, NASA may be read as one word or four individual letters.

To combat these potential issues, it is important to place the primary message first in the tweet, saving any at replies, hashtags, and links for the end of the tweet. This will ensure that the user will hear the message and remove any confusion (HowTo.gov 2012). Therefore a better version of the earlier tweet would be:

New social media book by @oaksquaresusan and @transitmom
http://tinyurl.com/bgcq3ov

Even though other websites such as Facebook do not have character limits or use special symbols such as hashtags, these tips can also be used. The primary message should always be at the beginning of the post. If sharing media, type an indication at the beginning to set the expectation for the rest of the message. This is usually [pic] for a picture, [video] for video, and [audio] for audio clips. A short message describing the media should then follow in addition to the link to the media or the media itself (HowTo. gov 2012).

Another strategy to make social media more accessible is specific for those who are deaf or hard of hearing: video captions and/or sign language translators. As websites such as YouTube and Vimeo become increasingly popular, people who have hearing deficits cannot always experience these videos. Since 2010, the Federal Communications Commission (FCC) has required broadcasters and video services to provide captions for online videos if they were first aired on television (Tsukayama 2012). This is an important ruling but it leaves a loophole for many popular videos on sites like YouTube, either because they were not first shown on television or they did not originate in America. The FCC rules do not apply to such videos.

Instead, for most popular online videos, supplying captions is up to the individuals who upload or own the videos. Many do not. However, the use of auto-caption technology does provide another source to allow the deaf and hard of hearing to experience these videos. YouTube provides auto-captions on most of its videos. However, because the technology is fairly new, it is not always accurate and, in some cases, is completely incorrect.

While auto-captioning is far from perfect, Google is continuously working to improve the technology for YouTube videos (Greenemeier 2011). At the moment, the best way to ensure that people who cannot hear can understand videos uploaded on social media is to provide captions. A transcript of a video can usually be uploaded along with the video.

A study of the use of captions in videos was conducted in 2009 by PLYmedia. The results indicated that adding captions to videos had a positive effect on viewership (eMarketer 2009). Not only did captions increase the total number of video views, but the addition of captions made it more likely viewers would watch the entire video. With all other factors controlled (i.e., video content), approximately 91% of the captioned video was viewed compared to 66% of a non-captioned video. The addition of captions benefits people with disabilities and can also improve accessibility for viewers who use a different first language. This research suggests that captions have an overall positive effect on videos and benefit people with disabilities and those with limited English proficiency.

Another tool that can be used for speeches, alerts, or videos is to include a sign language interpreter in a frame. During Hurricane Sandy, New York City Mayor Michael Bloomberg's news conference was recorded and uploaded online for residents to see and take precautions ahead of the storm. In the video, a sign language interpreter appeared next to the mayor to convey the message. While not everyone who is deaf or hard of hearing can read sign language, this approach allowed a wider range of residents to have access to what was then very critical information (Tsukayama 2012).

It is also incumbent upon a transit agency to supply information in an easy-to-understand format and make use of the many tools available to provide access to social media for those with disabilities. Composing easy-to-read tweets and providing captions on videos are just two of many easy ways to improve accessibility. Transit agencies can provide more accessible content if they understand how the information is presented; this may include experimenting with screen readers and auto-captioning software.

Conclusions

In summary, an equitable transportation system must provide all people the means to access information and express their opinions. This is not only a noble goal, it is also mandated in multiple U.S. laws. Transportation agencies must ensure that their use of social media includes appropriate practices, technologies, and supplementary traditional public involvement practices to ensure equitable decision making and information provision.

Although the digital divide was historically concerned with unequal access by lower income and non-white ethnic groups, these groups fare better with social media. However, older populations disproportionately use social media at a lower rate than younger people, creating a new digital divide. Similarly, some social media platforms may skew toward certain demographics. For example, LinkedIn has a substantial college-educated user base and Pinterest reveals a greater percentage of female users. Agencies must take all these biases into account and ensure that their outreach efforts include traditional means to supplement social media.

Agencies should also take into account non-English language populations on their blogs, tweets, and posts by translating into other common languages in their areas. Digital translation services are becoming more popular for less critical items, but the only way to make sure a social media piece is correctly translated is to use a human translator.

Finally, although technology such as screen readers and captioning can help people with visual or cognitive impairments, agencies must take into account how these tools are used when preparing social media tweets and posts by following best practices such as listing websites and hashtags last and including notes to explain that a video or photo has been attached. Through these practices, social media can be a useful tool for nearly all populations.

IS ANYBODY LISTENING? MEASURING IMPACTS OF SOCIAL MEDIA
Eric Rabe and Susit Dhakal

During last winter's blizzard, a city streets department turned to social media. By quickly posting on Facebook and Twitter, the department tried to get the word out about what streets were to be plowed first and the best routes to use for those who had to be outside. The department provided safety tips and even weather updates. When a power line fell and closed a major street, the location and detour routes were posted within minutes. But how many people actually saw the information? Did the effort really help the community deal with the snow? Were people paying attention?

———————

A big-city mass transit agency has worked hard to overcome an image of inefficiency and frequent delays. In recent years, internal measurements show that adding new buses and upgrading the rails significantly improved service. Training is better, and there is now a system-wide focus on customer service. The agency has tried to communicate those improvements using both traditional and social media, and there seems to be a lot of online conversation about the system. Are customers getting the message? Are there online supporters who are taking part in the discussion? Where are voices that might be willing to post positive comment?

———————

In a medium-sized city transportation department, two people, one full-time and one part-time, manage the online program. They have a lot to do. They

255

administer the website, post regularly on Facebook and Twitter, and write a blog. They also monitor social media, and they are the first to receive e-mails when users have questions or problems. Is the effort worthwhile in a time of tight budgets and fewer employees? Which of the social media tools is delivering the most for all the work? What indication does the department have that city residents are using these online channels?

Why Measure?

These sorts of questions confront managers of transportation agencies every day and the answers are hard to find. Often agencies wind up relying on anecdotes. Someone reports during a staff meeting that a negative comment was posted online, and it quickly becomes "accepted wisdom" that social media has turned ugly.

Resistance to measurement can be subtle. Executives sometimes feel they know how social media is working because they read the posts, tweets, and reader comments. Yet that effort may yield only a snapshot. At other times, the agency is unclear about exactly just what is to be measured. Readily available measures like number of "likes" on Facebook may not tell much. Practical issues such as staff and budget limits often get in the way. But there is more to the scant use of measurement than simply lack of time or inadequate direction.

The fact is that social media measurement is in its infancy. An executive at San Francisco Bay Area Rapid Transit (BART) frankly says, "Metrics for social media are not fully baked" (Moore 2012). Many automated online tools have limited capabilities. Services that devote significant human effort to measuring social media impact are often expensive, time consuming, and beyond the reach of government agencies.

Even so, transportation agency managers can explore the answers to questions like the ones above using tools that are available, easy to use, and frequently free. The only requirement is finding the time to use them and analyze what the tools reveal. That may depend upon how fully the agency recognizes the importance and the benefits of having a clear view of how well a social media program is performing.

Government Use of Social Media

When researchers at the Fels Institute of Government at the University of Pennsylvania asked 108 cities about measurement, 86% percent said they were trying to measure the effectiveness of their social media programs but had little time to do so and were not satisfied with the effort (Hansen-Flaschen and Parker 2012). Among 13 transit agencies polled by Fels in 2012, nearly 40% reported that they spent "little or no time" measuring the impacts of social media activities, and 85% said they spent less than 1 hour a week doing so. Yet some two thirds said they recognized value in measuring (Fels Institute 2012; Figure 6.12).

For several years, the Fels Institute has tracked government use of social media. In a 2009 report, *Making the Most of Social Media,* Fels provided government agencies with a road map to help them start using social media (Kingsley 2009). The Fels 2012 report, *The Rise of Social Government,* explored the use of social media at agencies in cities around the country and what best or promising practices had been discovered (Hansen-Flaschen and Parker 2012). The 2012 report showed that government makes extensive use of social media today. Nearly all of the 108 cities studied were using Facebook and Twitter to provide information and also to provide services to their communities (Figure 6.13).

Some services are well developed. Providing online notifications during an emergency is an example. When Hurricane Sandy hit the east coast of the U.S. on October 30, 2012, New York City's Office of Emergency Management used Twitter (@NotifyNYC) to let residents

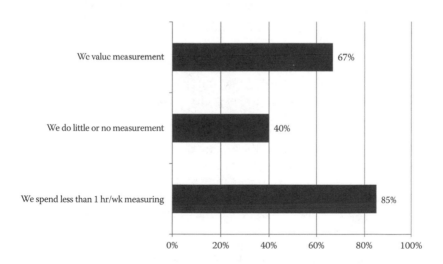

Figure 6.12 Value of social media measurement among transit agencies surveyed in 2012. (From: Fels Institute of Government. Survey of U.S. transit agencies conducted from September 17 through October 17, 2012. With permission.)

257

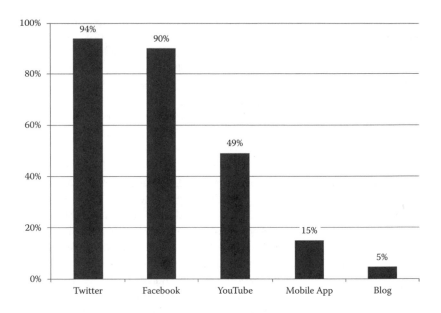

Figure 6.13 Twitter and Facebook use by local governments far exceeds use of other popular social media and mobile platforms as of August 2011. (From: Hansen-Flaschen, L. and K.P. Parker. 2012. *The Rise of Social Government.* Philadelphia: Fels Institute of Government. With permission.)

know about emergency services, food and water distribution sites, and other critical information (NYC OEM 2012). The Mayor's office relied on @ NYCMayorsOffice on Twitter to keep city residents updated and control rumors (2012). Figure 6.14 shows excerpts from the Twitter streams for both agencies.

Established services like these can save time both for government workers and constituents. Many government organizations monitor social media sites on which citizens may report problems and ask for help. The latest developments are attempts to generate constituent discussion of public policy questions such as economic development or policy feedback (Hansen-Flaschen and Parker 2012).

At the same time, cities rarely reported actually measuring the impacts of their social media work. As with the transit agencies surveyed, cities often focused on quantitative measures such as the number of friends on Facebook or the number of tweets issued by the city or department. Rarely do cities move to a qualitative measurement to distinguish

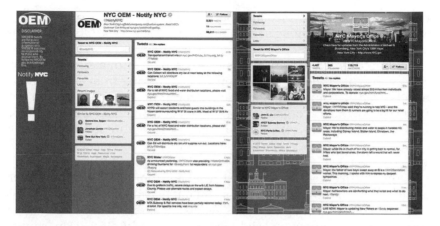

Figure 6.14 Tweets from New York's Office of Emergency Management and the Mayor's Office during recovery from Hurricane Sandy. (Twitter screen captures from @NotifyNYC and @NYCMayorsOffice)

between this kind of activity and the engagement or influence of social media users.

Return on Investment

When an agency, city, or department begins to work with social media, two questions quickly emerge. First, how does the social media program help to fulfill the mission or purpose of the agency? This relates to the second question: are the time and money spent on social media providing an acceptable "return on investment" (ROI)?

With competition for resources including staff time more intense than ever, government agencies must make choices. They need to understand whether their social media programs are worth the effort required. If the answer is, "not really," it's time to look for ways to have more impact or move on to something else. People in the transportation business do not have time to just spin their wheels.

The ROI can be considered in terms of improved efficiency and future costs avoided. Often agencies turn to social media because they hope that it will be cheaper and quicker to communicate with the public online than in alternative media including traditional news outlets and print. However, managers say that they often underestimate the amount of time social media can absorb. In the Fels research, cities rarely reported employment

259

of a full-time person devoted to social media, yet frequently staff noted the conflict between managing social media and other work. Often the social media work spilled over into free time in evenings or on weekends.

Future costs avoided are necessarily based on estimates. For example, future costs could include the expense of recovering from a crisis. If bad news is expected, an existing proactive social media campaign can help offset the blow to the organization's reputation and help it recover more quickly. Research shows that when brand and image are strong, they serve as an insurance policy against negative news and comment or future crisis (Lyon and Glen 2004).

Other future costs may be lost revenue from uneducated or disenchanted users and time and effort spent publicly defending agency decisions. The social media program is evaluated based on expenses such as these with and without a social media program in place. Departments can estimate the impact of user trends and project the impacts into the future. Past industry experience can help predict the costs of crisis recovery. Often agency strategic planners are doing this sort of work, but it may remain outside the sphere of the social media program. When the data are collected and applied to social media, they can be powerful in the ROI analysis. Table 6.2 shows some possible metrics for measuring return on investment.

To capture the ROI of a social media program, the frequency and quality of the comments in social media must be evaluated. Social media messaging can be evaluated, analyzed, and scored. Measurement should include the tone and how well key messages are reflected in reposts, retweets, and other online comments over time. Knowing the return on investment is a critical component to managing a social media effort, and it can show the importance of the social media program to an organization and its mission. However, to be most useful, the measurements should be

Table 6.2 Possible Measurements for Determining Return on Investment

Efficiency improvement	Future costs avoided
Electronic versus paper printing	Speedier crisis recovery
Electronic communication versus surface mail	Rider or user retention costs
Speed of delivery	Public support to limit advertising and public relations expenses
In-house production versus outside help	Complaint management costs

tied directly to operational objectives and priorities of an organization and to the goals of its communications and social media programs.

Continually Improving Effectiveness

It is also important to recognize that the social media landscape is continually changing. What works today may not work tomorrow as a fickle audience moves on to try new platforms or existing platforms mature to provide different options for users. Likewise, staff members become more confident and competent in their online abilities and the interests of organizations change over time, offering new capabilities and needs.

Agencies must evaluate all these factors to ensure that they are as effective as possible in reaching those they want to reach and truly engaging them in discussions that provide useful information, inform the agency about new opportunities and needs, and draw public support for services, plans, and programs. A social media measurement program will help organizations stay on top of the online evolution.

Deciding What to Measure

Just what sort of information is most useful in evaluating a social media program? Where can organizations find it and what do they do when they have reliable information? Setting goals for social media programs is key to their success. Measurement is best aimed at determining how well the goals of a social media program are being met.

If a social media program seeks to increase rider awareness of transit options, a transit agency should measure the change in rider awareness and what caused the change. Part of the answer may lie in measuring changes in ridership but it is also important to measure the contribution of social media to new awareness. How did riders learn about changes? Do they rely on social media for up-to-date information?

If a social media program's goal is to improve the public image of a transportation department, then the change in the image and the reason for the change should be measured, probably through a survey. Also measuring online comments, often conveyed in real time and directed to acquaintances of writers, can give insight into how well an agency or department is perceived. Indeed, measuring these comments may reveal writers' thinking in a way that surveys and focus groups cannot.

The goal of social media activity may be to improve operations. The effectiveness of operations improvements and social media contribution to the improvements can be measured. For example, perhaps a transit organization has created a new system that permits riders to enter information about planned trips and then calculates the right combination of transit options to get riders to their destinations quickly. The new system will be promoted primarily through social media. What will spell the success of such a program? Transit administrators may want to know how often readers of social media use the new trip planner, but also how many use it more than once, how many tell their friends about it, and what sorts of comments users make about their experiences to the agency and to their online friends. Measurement is invaluable for determining whether a new system is understood and effective and whether riders like the experience.

Why Just Counting Friends Is Not Enough

When measuring social media, it is critical to measure activities and also measure engagement and influence of those using sites and reading posts. Many simple activities occur almost automatically on social media, and just counting connections does not indicate the strength of a relationship.

With the flick of a finger, someone can decide to "follow" an agency on Twitter or "like" it on Facebook. While each of those actions may be better than nothing, neither indicates strong support or a strong opinion. A Facebook friend may never return. A Twitter follower may simply skip over an agency's posts while looking for tweets from relatives, schoolmates, or even celebrities. Worse, simply counting activities may be wildly misleading. An agency could be getting hundreds of mentions on Twitter that may seem like good news, but without measuring the tone of the comments, the organization will not know whether it is being praised or trashed.

One transit agency told Fels Institute researchers that the agency put little stock in Facebook "likes," that are considered popular measures of success on the social media site (Kenton 2012). Agency officials believe it is more meaningful to measure the engagement signified when a Facebook user *shares* a link to an agency page or posting. When that happens, the user puts his or her personal imprimatur on the agency's material and is saying, "Friends, I've found something I think is really useful or interesting, funny or entertaining, and I want to share it with you." If the user has a large base of followers, the influence combined with a positive tone may be quite significant for an organization.

Measurement that emphasizes actions shows this engagement most clearly. Some activities, such as responses posted on a site, may be easy to measure. Twitter allows users to gather information about retweets when another Twitter user forwards an original message to a new group of Twitter users. The 2012 Fels report (Hansen-Flaschen and Parker) finds this example:

> For Philadelphia's Desiree Peterkin-Bell, tracking the reach of retweets highlights social media's power of garnering the city national attention overnight: "I track who retweets our posts. Once Questlove and the White House retweeted a post we sent out, so that reached several million more people."

That single retweet showed strong engagement and produced a big influence online.

Quick searches can determine how often a link from an agency's material is forwarded by a reader, retweeted, or posted on a Facebook page. Researchers know that there is more value in some reader actions than in others (Keenan 2012). In evaluating the impact of social media, it is important to take note of the different sorts of actions and assign judgments of what they say about user involvement.

What Do Cities Measure?

Both the Fels *Rise of Social Government* study and follow-up research show that transit agencies and other government units frequently do not measure or are just beginning to measure the impacts of their social media programs. For example, *The Rise of Social Government* reported that only 52% of cities and agencies even attempted to quantify social media activity as part of an overall reporting process. However, 84% expect this to become a future requirement for their teams (Hansen-Flaschen and Parker 2012).

The Rise of Social Government notes that cities today frequently rely on counting things—number of followers, number of posts or comments, number of YouTube views, number of shares, reposts, and retweets. Less common is rating the tone and reach of the individual posts, comments, and retweets.

Overview of Current Measurement Tools

Smart government agencies are focusing their measurement on engagement and impact. Their questions are about activities by social media users that indicate some involvement beyond simply clicking the "like" or "follow" button on Facebook and Twitter, respectively.

The big social media sites are making more and more information available. Twitter is beginning to offer counts of retweets of posts, and free online tools such as HootSuite allow users to track actual retweets. Facebook Insights offers a range of statistics related to an individual page. Of course, for a large agency with an active social media presence, reviewing online comments can be time consuming. Furthermore, those statistics may not reach the core of social media engagement and influence. Creative social media managers are asking questions such as:

- Can we now participate in conversations in which we had no voice in the past?
- Are we having a dialogue with constituents?
- Is our perspective influencing the overall conversation?

Although capabilities are growing, most tools now available on the web are limited in their ability to answer questions like these that relate to engagement and influence. Obtaining deeper insight requires an even-handed evaluation of the tone (positive, negative, or neutral) of a comment and the number of readers who may have seen it (reach) either by reading the original post or discovering it through links or other online comment.

Monitoring and Measuring

Certainly one of the key functions of a social media program is monitoring or listening. Measuring the impact flows from monitoring, but measuring is an additional and separate activity.

Monitoring

Even before a social media program begins, it is important to determine what conversations are taking place about an agency, its service, its personnel, and its mission. With millions of sources participating in thousands of conversations, how can anyone sort through it all?

The Rise of Social Government notes that 95% of all cities monitor social media regularly, with two thirds monitoring on a real-time basis

> **Monitoring** is the act of listening. It is the practice of observing the sharing of information online to determine who is saying what to whom. Monitoring may also offer a first indication of trends, problems to be resolved, or even crisis.
> **Measuring** is assessing what is said and making a judgment about the impact, positive or negative, on the objectives set by an organization's social media program. It is a regular ongoing process.

or hourly (Hansen-Flaschen and Parker 2012). Like cities, transportation agencies studied by Fels (2012) monitor social media primarily to understand what is being said. Additionally, responding to an online conversation can be an effective way to attract new followers and potential supporters.

Monitoring also can help defuse public concerns or frustrations by giving an agency the opportunity to weigh in before overwhelming tension builds. Monitoring may also provide early warning of issues facing a department or even a first alert to a problem or crisis in the system. Figure 6.15 shows how much time transit agencies devote to monitoring

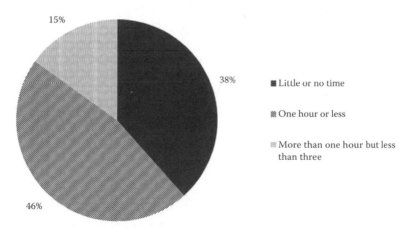

Figure 6.15 Most transit agencies surveyed by Fels reported spending less than 60 minutes per week monitoring or measuring social media. None reported spending more than 3 hours per week on such activities. (From: Fels Institute of Government. Survey of U.S. transit agencies conducted from September 17 through October 17, 2012. With permission).

social media. As Hansen-Flaschen and Parker (2012) reported, cities monitor social media elements including:

- Public posts that mention the city and the demographics behind them.
- Popular and "trending" topics to determine the overall drift of the online conversation.
- Keywords developed to capture conversations that may be important. A streets department may search for the "pothole" keyword, for example.
- Other organizations related to the city or that monitor or follow city activities.

Many tools are available online to help agencies monitor conversations. Some are built into specific social media platforms, including Facebook Insights and YouTube Insight. Others are available from outside parties, including Google Analytics, HootSuite, and TweetDeck (the latter two for Twitter). Each provides a way to judge the activities on web pages or social media accounts; all are free and easy to use. Other useful monitoring tools are listed in Table 6.3. The inventory of available tools may change in response to market forces and technical requirements; this list presents a selection of tools available as of this writing.

Measuring
While monitoring may be a first step in measuring the impact of social media, an effective social media measurement program provides important insights not available from simply listening to social media. At the start, a serious measurement strategy must assure a continual process. Only by using a disciplined approach will agencies be able to gather data that reveal trends showing progress and success over time.

Developing Measurement Programs

It is most effective and probably easiest to develop a measurement program at the same time an agency begins its social media program. Teams can discuss not only how to manage the work of social media, but also two other important considerations: (1) the overall objectives of the program and (2) the measurements that will show progress toward reaching those objectives. Careful planning and a clear statement of objectives and social media's ability to help reach those objectives are essential from the outset.

Table 6.3 Online Monitoring and Measurement Tools

Online Monitoring Tools

Addict-o-matic: After users enter a topic of interest, site instantly provides links to all the latest online content across all major web platforms (Bing, Yahoo, YouTube, GoogleBlog, WordPress, etc). http://addictomatic.com/

BlogPulse: Follows and reports trends in blog posts. www.blogpulse.com

Facebook Insights: Provides metrics and trends for consumption, creation, interaction and growth on administrator's Facebook page. https://www.facebook.com/help/search/?q=insight

HootSuite: Social media dashboard to measure and manage social networks. https://hootsuite.com/

Monitor.us: All-in-one web server monitoring service (useful for IT and web developers). http://www.monitor.us/index.jsp

Social Mention: Real-time social media search and analytics platform aggregating all online user-generated content into a single information stream. http://www.socialmention.com

TweetDeck: Personal real-time browser to connect with user's online contacts across all social media platforms. http://tweetdeck.com/

Online Measurement Tools

TweetReach: Within seconds users can learn how many people viewed a particular tweet. http://tweetreach.com

Google Analytics: Provides detailed insights into website traffic and marketing effectiveness. http://www.google.com/analytics

Sitetrail: Website analysis platform. Along with social media analysis, site offers free analysis of links, traffic, and visitors. http://www.sitetrail.com

Netvibes: All-in-one dashboard to monitor social media accounts and other web-based applications. http://www.netvibes.com

Radian6: Fee-based social media analysis services plus free resources and support information. http://www.salesforcemarketingcloud.com/

Facebook Insights: Metrics for Facebook pages. https://www.facebook.com/help/pages/insights

Organizations may have a more general communication plan already in place. If so, that's a logical place to begin considering the objectives of a social media program. What's called for is a clear view of what logically can be accomplished rather than wide-eyed optimism.

The IT department may be able to help. Karen Hale, the communications director in Salt Lake City, Utah, took that approach. Her communications team turned to the IT department for help finding handy useful tools, help in streamlining ways to use them, and help in incorporating them into existing measurement schemes (Hansen-Flaschen and Parker 2012). Developing a measurement program has three steps.

DEVELOPING A MEASUREMENT PLAN

1. Clearly state the objective of the social media program, including expected actions by the target audience.
2. Determine where important online conversations are taking place.
3. Analyze data gathered online in light of objectives.

First Step: Articulate Program Objectives

One step that every PR professional knows is important and nonetheless is frequently ignored in the social media world is stating objectives clearly. Often social media programs begin with one or perhaps several enthusiasts inside an organization who evangelize and may even have started blogs, websites, Facebook pages, and Twitter feeds on their own initiatives. Sooner or later these spontaneous social media efforts are corralled by the larger organization, coordinated, and streamlined. But often little thought is given to what the organization actually wants to accomplish and how social media activity relates to the overall mission of the department.

Is the goal to improve the reputation of the agency or get more riders on the transit system? Does the agency want to hear from drivers about the next road project or encourage bicycle commuting? Each of these could be a useful objective of a transportation agency and may be incorporated into communications goals that in turn must be tied to the overall goals of the agency. Communications goals may answer questions such as:

- **Reach:** How many people are in the target audience and how many see agency communications?
- **Understanding:** How many of those reached comprehend new transit programs or ideas and act on that information?
- **Tone:** Is online comment about the department positive, negative, or neutral?
- **Advocacy:** How many are advocating on behalf of the agency and telling their networks the story?

Choosing specific goals early on will help determine which social media channels are used, what the voice of each is to be, and what is expected of each channel. If the idea is to build a community of bike riders, Facebook may be the best answer. If an agency wants a real-time channel to reach drivers in an emergency, Twitter might work better. A social media objective can be any issue that makes sense to an organization or agency. A short list may apply to transportation organizations:

- Increase transparency of the agency.
- Give users real-time information about changes and improvements.
- Gather input from users about their experiences.
- Improve customer service and respond to user issues like potholes.
- Start users talking about improvements or changes that could make the system better.

While it is important to clearly define what an agency hopes to accomplish with the new effort, it is just as important to consider how to draw followers, members, and friends to the new social media conversation. After determining the overall objective of a social media program, it is important to articulate that objective and get agreement within the organization that the choice is the right one. That puts the program on solid ground when measurements are selected to determine whether the department's objective is being achieved.

Second Step: Find Conversations

Transportation agencies can be sure that someone is almost certainly discussing them in the social space right now. Even if the agency is not yet part of that discussion, it is important to follow it after a social media program is created. In establishing a social media program, leaders must determine which conversations are the most important given the agency's objectives. A week or two spent searching for important keywords, monitoring discussions online, and tracking mentions of the agency or priority initiatives can provide plenty of information about online places where the agency should be involved in the conversation.

A transportation department may be part of a larger government entity. What conversations taking place on the city or state online sites focus on transportation issues? Citizen neighborhood sites are increasingly popular. Is it worth the time required to participate in dozens of these that may exist in an urban area? Perhaps the department plans to establish a new Twitter stream or Facebook page. After some careful monitoring, the team may decide time is better spent engaging in conversations that are already online—existing sites, the sites of the larger government entity, or sites run by entirely independent groups. Monitoring social media allows an agency to determine where this type of communication can be effective based on its objectives.

Third Step: Start Measuring

A social media measurement program must be a regular and ongoing effort. The set of issues to be studied must be consistent over time. Measurements should track three important elements of communication:

- **Exposure:** How many people are reached? Agencies will want to track increases in followers or visitors and should compare month-to-month growth rates. Did a particular event drive a spike in growth?
- **Engagement:** Measure interactions with material posted online. Track comments, forwards, and references on sites the agency does not control.
- **Influence:** The extent to which agency-created material influences a larger conversation.

This requires rating the *quality* of the comment, repost, retweet, or other response. Generally, agencies can rate comments by reading and scoring the material based on tone (positive, negative, or neutral) and reach of the response (broad, moderate, or narrow). The result can provide a much richer picture of social media impact than is possible without taking this step.

In interviews with transit agencies around the country, Fels (2012) researchers found none that devoted more than 3 hours a week to measurement. Most spent less than 90 minutes. With so little time for social media measurement, much of the rich information about the effectiveness of the social media program will continue to lie undiscovered beneath the surface of the web.

Quantitative and Qualitative Analysis

Measuring is both quantitative and qualitative. *Quantitative* measures such as the numbers of followers, comments, retweets, and posts are fairly easy to find. As discussed earlier, most social media platforms including Facebook and Twitter provide the numbers as a matter of routine. On the other hand, *qualitative* measures of engagement and influence involve making human judgments about impact, tone, and value of conversations. That means more work, but the deeper insights from qualitative measurement are essential to judge the true importance of social media interactions accurately.

It may be that the team has decided that an active presence on Facebook will help spread the word about what a transit agency is doing to improve service. The team may decide that a quantitative assessment (counting followers) is important. However, a deeper qualitative analysis can give insight into the tone of comments of followers, their reach (how many friends they have and regularly contact), the key users who return more than two or three times, and what they are saying.

That can lead to a strategy for more intensely engaging key users, perhaps responding to their comments directly on a Facebook page or elsewhere. Tracking reposts or links to the Facebook page will begin to give the team a sense of the influence of key users and help identify other online conversations that may be relevant. Tracking the growth of total followers and also *engaged* followers are indicators of success in getting an agency's message out.

Role of Lead Measurement Expert

Successful programs require a lead person or expert to be responsible for monitoring and measuring. If this responsibility is not in the hands of an accountable staff member, measurement can become an afterthought, leaving social media programs operating on intuition and instinct rather than sound judgment. The lead measurement expert has several important responsibilities.

- Overall responsibility for the measurement program. He or she should be responsible for monitoring and measuring and must ensure that the measurement program is focused and disciplined.
- Developing measurement metrics at the outset. The management team must agree to the metrics and they will have to be adjusted over time. Some trial and error will be required to devise metrics that are both manageable and effective. As the social media program develops, the metrics will be adjusted.
- Regular analysis, at least monthly (weekly is better), of the measurements and the program's success in reaching the established objectives. If the social media program seems to be faltering, this individual is responsible for sounding the alarm.
- Ability to serve as the team's expert in the use of measurement tools. She or he will be familiar with the available tools and understand which tools work best for the agency's program.

271

Getting Creative

In developing a social media measurement program, creativity may be the most important element. The team must ask, "What will help us know whether we are having a significant impact on our objectives?" When measurement is linked closely to established objectives, the team will develop ideas for measurements that are unique and tied to results that matter.

The Econsultancy marketing website identified several dozen metrics that social media users can utilize (Lake 2009). The metrics include many quantitative measures that emphasize engagement and impact. For instance, Econsultancy suggests tracking the number of pages printed by users or the time users spend on key pages. Those are signals that a user was more than a casual visitor to a Facebook post or website.

The Southeastern Pennsylvania Transportation Authority (SEPTA) in Philadelphia monitors social media to be sure complaints, requests for information, and service issues are handled quickly and efficiently when they appear on social media. SEPTA measures both social media activity and also the follow-up responses to requests and complaints, "... very similarly to how our traditional customer contacts are measured: ... time to resolution, percent engaged, customers' satisfaction and internal efficiency" (Busch 2012).

San Francisco Bay Area Rapid Transit (BART) considers its website the hub of online activity and "... the focus of calls to action in all of our channels: e-mail, text, Twitter, Facebook, etc. Tracking website referrals is a great way to compare how social media performs versus other tools like e-mail" (Moore 2012). BART's analysis of digital interactions shows that social media sites like Facebook and Twitter account for far fewer customer interactions than e-mail and website activity, and thus it puts more emphasis on the latter.

The Utah Transit Authority uses Twitter heavily to engage riders and stages occasional "Twitter chats." Staff reported, "We keep track of conversations beyond retweeting. We diligently track each day how many accounts are talking about us regardless of whether or not they '@-mention' us or talk with us directly. We also measure the raw number of clicks for the links we put up on Twitter" (Shubin 2012). These measurements help determine whether the effort made on Twitter is paying off.

The Massachusetts Bay Transportation Authority uses social media to reach mainstream news media. Staff members post content on Twitter and then monitor click-throughs to the content and to retweets. "Retweets just show that a message has been passed along; clicks show interest," a spokesman said. "We often use the interest measure (determined by counting clicks) to help calibrate stories that we push to more mainstream media outlets" such as newspapers or broadcast news reporters in the hopes of traditional media coverage (Robin 2012).

Measurement in 180 Minutes per Week

Agencies convinced of the value of social media measurement find the time to measure. However, few of the agencies that Fels contacted were able or willing to spend more than a little time each week measuring the impact of social media. How can agencies obtain meaningful measures by dedicating only 3 hours a week?

Limited measure time will not allow agencies to pore over every tweet or online comment. Instead, they will need to select a sample from their online activity for deeper analysis. However, they can use tools to identify a pool of comments that they can monitor and draw from for further study. One recommended approach to measurement and analysis is to (1) set aside a specific time each week for measurement activity; and (2) start by creating automatic web reports using online tools like those listed below. Agencies can choose the three or four tools that seem most appropriate based on their social media program and its objectives.

If a **web page and blog** are the primary social media channels, agencies can look at Google Analytics[*] for analysis and search or aggregation tools to capture web-based discussions about the agency and service. Another monitoring tool worth exploring is monitor.us[†].

If **Facebook** is a primary channel, agencies can check out Facebook Insights (https://www.facebook.com/help/pages/insights) to track activity on the agency Facebook page or pages.

For **Twitter** posts, chats, and other conversation, HootSuite[‡] and TweetDeck[§] are tools to monitor and track activity. Each requires some setup, but after that they provide streams of continuous data.

Successful social media practitioners can recruit a team of two or three associates to read online comment and evaluate its impact. Meeting once a week to score a sample of randomly selected content is an easy and convenient way to judge results. A standard report can be completed

---

[*] http://www.google.com/analytics
[†] http://www.monitor.us/website-monitoring
[‡] https://hootsuite.com/
[§] http://tweetdeck.com/

easily and quickly from web-based monitoring and the qualitative input of the agency team.

Using the same report and metrics each week will help keep the job manageable. Over time, trends become evident and an agency can make judgments based on observed activity.

Of course, measurement can be as simple or as elaborate as imagination and time (and budget) allow. The agencies surveyed by Fels researchers relegated measurement to 60 minutes a week or less and agreed that the more time spent, the better the results. Three hours is a good initial commitment. By spending more time, social media teams will be rewarded with deeper and more insightful results. One approach is to start only with information that can be managed reasonably. As the team works more and more with measurement tools that are available online, members will become more comfortable with them and begin to gather important data in less time.

Measurements should be directed at determining how well a social media program meets an agency's objectives. A great deal more may be learned but measured results tied to the agreed-upon objectives serve as the gold at the end of the social media rainbow. Those results will add insight to social media work and insights that will make programs more effective and stronger.

For agencies that can afford to get help with their social media measurement programs, a number of organizations and online tools will help for a fee. Some are expensive and worth it, but may be beyond the means of agencies. However, many excellent options are available. Table 6.3 lists some online measurement tools.

REFERENCES

Air Force Public Affairs Agency. http://www.af.mil/shared/media/document/afd-091210-037.pdf

American Association of State Highway and Transportation Officials (AASHTO). 2012. Third Annual State DOT Social Media Survey.

American Council for Technology–Industry Advisory Council. 2011. Best Practices Study of Social Media Records Policies. http://www.actgov.org/knowledgebank/whitepapers/Documents/Shared%20Interest%20Groups/Collaboration%20and%20Transformation%20SIG/Best%20Practices%20of%20Social%20Media%20Records%20Policies%20-%20CT%20SIG%20-%2003-31-11%20%283%29.pdf

Apple. 2013. iPhone Vision. *Apple.* http://www.apple.com/accessibility/iphone/vision.html

Brault, M.W. 2012. Americans with Disabilities: 2010. United States Census Bureau. http://www.census.gov/prod/2012pubs/p70-131.pdf

Bregman, S. 2012. *Uses of Social Media in Public Transportation.* TCRP Synthesis Report 99. Washington, D.C.: Transportation Research Board.

Brown, L. 2012. Social media in the public involvement process? Yes, but be specific and have a plan. *Talking Transportation.* http://talkingtransportation.wordpress.com/2012/06/26/social-media-in-the-public-involvement-process-yes-but-be-specific-and-have-a-plan/

Bullard, R.D. 2003. Addressing urban transportation equity in the United States. *Fordham Urban Law Journal* 31, 5.

Busch, A., Southeastern Pennsylvania Transportation Authority, e-mail message to author, October 4, 2012.

Butler, P. 2012. Disability activists use social media to put care cuts on the political agenda. *The Guardian,* http://www.guardian.co.uk/society/2012/aug/20/disability-activists-media-care-cuts

Center for Internet & Society (CIS). 2012. Making Mobile Phones and Services Accessible for Persons with Disabilities, G3ict and ITU, August.

City of Calgary. 2011. Transit trouble takes to Twitter. *CTV News,* July 19. http://calgary.ctvnews.ca/transit-trouble-takes-to-twitter-1.672477

City of New York. 2011. Social Media Customer Use Policy. http://www.nyc.gov/html/misc/html/social_media_policy.html

City of Seattle. 2009a. Legislation, Policies and Standards: City of Seattle Facebook Standard. http://www.seattle.gov/pan/SocialMedia_Facebook.htm

City of Seattle. 2009b. Legislation, Policies, and Standards: City of Seattle Twitter Standard http://www.seattle.gov/pan/SocialMedia_Twitter.htm

Desantis, N. 2011. The social subway. *Washington City Paper,* April 12. http://www.washingtoncitypaper.com/articles/41336/wmata-twitter-guru-dan-stessel-wants-dc-to-love-metro/full/

Duggan, M. and J. Brenner. 2013. The demographics of social media users, 2012. Pew Internet & American Life Project. http://pewinternet.org/Reports/2013/Social-media-users.aspx

Einhorn, A.B. 2012. 7 Reasons every government agency needs a social media policy. *OhMyGov.* http://blog.ohmygov.com/blogs/general_news/archive/2012/07/25/7-reasons-every-government-agency-needs-a-social-media-policy.aspx

El Pasajero tweet. https://twitter.com/elpasajero/status/304643974302404608

eMarketer. 2009. Online video subtitles (Duh!). http://www.emarketer.com/(X(1)S(33jocm55r4utiaqaee33hvy2))/Article/Online-Video-Subtitles-Duh/1007004

Falls, J. 2012. The problem with engagement. *Social Media Explorer.* http://www.socialmediaexplorer.com/social-media-marketing/the-problem-with-engagement/

Federal Communications Commission. 2013. Freedom of Information Act, Section 552(b). http://transition.fcc.gov/foia/5USC552b.pdf

Federal Highway Administration. 2011. FHWA Social Media/Web 2.0 Management. http://www.fhwa.dot.gov/legsregs/directives/orders/137014.htm

Fels Institute of Government. 2012. Survey of U.S. Transit Agencies. September 17 through October 17.

Fox, S. and J.L. Boyles. 2012. Disability in the digital age. Pew Internet & American Life Project. http://www.pewinternet.org/Presentations/2012/Aug/Disability-in-the-Digital-Age.aspx

Google. 2013. Google Display Network Ad Planner: Pinterest.com. https://www.google.com/adplanner/#siteDetails?uid = pinterest.com&geo = US

GovDelivery. Government, Facebook and Twitter: Best Practices Brief. http://www.govdelivery.com/pdfs/Social_Media_Best_Practices.pdf

Government Accountability Office. 2005. Better dissemination and oversight of DOT's guidance could lead to improved access for limited English-proficient populations. http://www.gao.gov/products/GAO-06-52

Government Digital Service. 2012. Social Media Guidance for Civil Servants. Part 1: Guidance on the Use of Social Media. http://www.cabinetoffice.gov.uk/sites/default/files/resources/Social_Media_Guidance.pdf

Greenemeier, L. 2011. Say what? Google works to improve YouTube Auto-Captions for the deaf. *Scientific American.* http://www.scientificamerican.com/article.cfm?id = google-youtube-auto-caption

Hansen-Flaschen, L. and K.P. Parker. 2012. The rise of social government. Fels Institute of Government, The University of Pennsylvania. http://www.fels.upenn.edu/news/rise-social-media-local-government

Hollier, S. 2011. Sociability: Social media for people with a disability. *Media Access Australia.* http://www.mediaaccess.org.au/sites/default/files/files/MAA 2657-%20Report-OnlineVersion.pdf

HowTo.gov. 2012. Making social media more accessible. http://www.youtube.com/watch?feature = player_embedded&v = aMlFWIu6rpY#!

Hrdinova, J., N. Helbig, and C.S. Peters. 2010. *Designing Social Media Policy for Government: Eight Essential Elements.* Albany: Center for Technology in Government, State University of New York. http://www.ctg.albany.edu/publications/guides/social_media_policy/social_media_policy.pdf

Indiana Department of Transportation (INDOT). n.d. Connect with social media. http://www.in.gov/indot/3074.htm

Kapanci, S. and J. Grant. 2012. Disability and social media. http://www.slideshare.net/SezenKapanc/disability-social-media-15379803

Kaufman, S.M. 2012a. How social media moves New York: 1. Twitter use by transportation providers in the New York region. Rudin Center for Transportation, Wagner School of Public Service at New York University.

Kaufman, S.M. 2012b. How social media moves New York: 2. Recommended social media policy for transportation providers. Rudin Center for Transportation, Wagner School of Public Service at New York University.

Keenan, E. 2012. 4 Social media evaluation metrics more important than likes. *Digital Marketing Blog, SiliconCloud.* http://www.siliconcloud.com/blog/bid/90934/4-Social-Media-Evaluation-Metrics-More-Important-than-Likes

Kenton, R.L. New York City Department of Transportation, e-mail message to author, September 19, 2012.

King County (Washington State). 2010. Social Media Guidelines. http://www.kingcounty.gov/exec/~/media/exec/socialmedia/documents/Social_Media_Handbook_12_6.ashx

Kingsley, C. 2009. Making the most of social media. Fels Institute of Government, The University of Pennsylvania. http://www.fels.upenn.edu/news/making-most-social-media

Lake, C. 2009. 35 Social media KPIs help measure engagement. *Econsultancy*, http://econsultancy.com/us/blog/4887-35-social-media-kpis-to-help-measure-engagement

Litman, T. 2010. Evaluating transportation equity: guidance for incorporating distributional impacts in transportation planning. Victoria Transport Policy Institute.

Lyon, L. and T.C. Glen. 2004. A relational approach examining the interplay of prior reputation and immediate response to a crisis. *Journal of Public Relations Research*, 16, 213–241.

Madden, M. and K. Zickuhr. 2011. Sixty-five percent of online adults use social networking sites. Pew Internet & American Life Project. http://pewinternet.org/Reports/2011/Social-Networking-Sites.aspx

Market Connections. 2012. Federal media and marketing study overview. http://www.marketconnectionsinc.com/Reports/federal-media-a-marketing-study-2012.html

Market Connections. 2011. Social media in the public sector. http://www.marketconnectionsinc.com/Reports/social-media-in-the-public-sector-2011.html

Massachusetts Department of Transportation. Social Media Use and Policies. http://www.massdot.state.ma.us/main/MassDOTSocialMedia.aspx

Minnesota Department of Transportation (MNDOT). 2011. Mn/DOT Policy Position Statement. http://www.ttap.colostate.edu/downloads/clearinghouse/MNDOT_SOCIAL.pdf

Missouri Department of Transportation (MoDOT). n.d. Social Networking Guide lines for Employees. http://www.lifesaversconference.org/workshophandouts2011/Holloway.pdf

Moore, T. Bay Area Rapid Transit (BART), e-mail message to author, September 27, 2012.

National Archives and Records Administration. 2010. Guidance on Managing Records in Web 2.0: Social Media Platforms. NARA Bulletin 2011-02. http://www.archives.gov/records-mgmt/bulletins/2011/2011-02.html

National Council on Disability. 2011. The Power of Digital Inclusion: Technology's Impact on Employment and Opportunities for People with Disabilities. http://www.ncd.gov/publications/2011/Oct042011

National League of Cities Risk Information Sharing Consortium. 2011. Building Local Government Social Media Policies. http://www.nlc.org/File%20Library/Utility%20Navigation/About%20NLC/SML/NLC-RISC/RISC-2011-Social-Media-Policies.pdf

NCAM. 2012. Researching accessibility gaps in transit hub communication systems and standards. http://ncam.wgbh.org/experience_learn/media_home_work/ncat

New Jersey Transit Facebook page. http://www.facebook.com/NJTRANSIT/posts/375854222494854

New York City Mayor's Office. 2012. @NYCMayorsOffice. *Twitter*. https://twitter.com/NYCMayorsOffice

New York City Office of Emergency Management (NYC OEM). 2012. @NotifyNYC. *Twitter*. https://twitter.com/NotifyNYC

New Zealand Department of Internal Affairs. 2012. *Social Media in Government: How to Handle a Mishap*. http://webtoolkit.govt.nz/files/Social-Media-in-Government-How-to-Handle-a-Mishap-v1-0.pdf

NextStop STL. 2012. Privacy. Metro Transit St. Louis. http://www.nextstopstl.org/about/policies/privacy

NM Incite. 2012. State of Social Customer Service 2012. http://nmincite.com/wp-content/uploads/2012/10/NM-Incite-Report-The-State-of-Social-Customer-Service-2012.pdf

NTIA. 1995. Falling through the net: a survey of the "have nots" in rural and urban America. http://www.ntia.doc.gov/ntiahome/fallingthru.html

Pelgrin, W.F. 2010. *Cyber Security Guideline G10-001: Secure Use of Social Media*. Albany: New York State Office of Cyber Security and Critical Infrastructure Coordination. http://www.empire-20.ny.gov/sites/default/documents/cyber.pdf

PRWeb. 2012. North Carolina unveils cutting edge social media archive using ArchiveSocial. http://www.prweb.com/releases/archivesocial/social-media-archiving/prweb10197203.htm

Rainie, L., J. Brenner, and K. Purcell. 2012. Photos and videos as social currency online. Pew Internet & American Life Project. http://pewinternet.org/Reports/2012/Online-Pictures.aspx

Reform Support Network. 2012. Building enduring race to the top education reforms: using social media to engage with and communicate to key stakeholders. http://www2.ed.gov/about/inits/ed/implementation-support-unit/tech-assist/using-social-media-pub.pdf

Rivera, R. 2012. Social media strategy was crucial as transit agencies coped with hurricane. *The New York Times*. http://www.nytimes.com/2012/12/15/nyregion/social-media-strategy-crucial-for-transit-agencies-after-storm.html?_r = 1&

Robin, J.K., Massachusetts Bay Transit Authority, e-mail message to author, September 27, 2012.

Shubin, E., Public Relations and Marketing Manager, Utah Transit Authority, e-mail message to author, September 18, 2012.

Skjekkeland, A. 2009. Fast records management (#ERM) facts from new AIIM survey. *GovLoop.com*. http://www.govloop.com/profiles/blogs/fast-records-management-erm

Smith, A. 2010. Mobile access 2010. Pew Internet & American Life Project. http://www.pewinternet.org/Reports/2010/Mobile-Access-2010.aspx

Smith, A. and J. Brenner. 2012. Twitter use 2012. Pew Internet & American Life Project. http://pewinternet.org/Reports/2012/Twitter-Use-2012.aspx

Society for New Communications Research. 2007. Best Practices for Developing and Implementing a Social Media Policy. http://sncr.org/sites/default/files/documents/tip%20sheets/SNCR_Social_Media_Policy_Best_Practices.pdf

Solis, B. 2010. *Engage! The Complete Guide for Brands and Businesses to Build, Cultivate, and Measure Success in the New Web*. New York: John Wiley & Sons.

State of North Carolina, Office of the Governor. 2009. Best Practices for Social Media Usage in North Carolina. http://www.records.ncdcr.gov/guides/best_practices_socialmedia_usage_20091217.pdf

State of Utah. 2009. Social Media Guidelines 4300-0029. http://www.utahta.wikispaces.net/file/view/State+of+Utah+Social+Media+Guidelines+9.29.pdf/177053623/State%20of%20Utah%20Social%20Media%20Guidelines%209.29.pdf

State of Washington, Office of the Governor. 2010. Guidelines and Best Practices for Social Media Use in Washington State. http://www.governor.wa.gov/media/guidelines.pdf

TBARTA. https://twitter.com/TBARTA/status/58523832876675072)

Texas DOT website. http://www.dot.state.tx.us/public_involvement/social_media.htm

Tsukayama, H. 2012. Growth of viral video leaves deaf in the dark. *Washington Post*. http://www.washingtonpost.com/business/technology/growth-of-online-video-leaves-deaf-community-in-the-dark/2012/11/28/4048e4ac-389c-11e2-8a97-363b0f9a0ab3_story.html

U.S. Air Force Public Affairs Agency. 2009. Air Force Web Posting Response Assessment. http://www.af.mil/shared/media/document/afd-091210-037.pdf

U.S. Department of Justice. 2011. What is FOIA? http://www.foia.gov/about.html

U.S. Environmental Protection Agency. 2012. EPA Environmental Justice Basic Information. http://www.epa.gov/environmentaljustice/basics/ejbackground.html

U.S. General Services Administration. 2012. Top Ten Best Practices for Multilingual Websites. http://www.howto.gov/web-content/multilingual/best-practices

Waldman, L. 2011. Social media discovery: the City of New York chooses Nextpoint. *Nextpoint Company Blog*. http://www.nextpoint.com/releases/3470/social-media-discovery-the-city-of-new-york-chooses-nextpoint/

Washington State Legislature. Undated. Record Retention Schedules: Destruction and Disposition of Official Public Records, Office Files and Memoranda. RCW 40.14.060. http://apps.leg.wa.gov/RCW/default.aspx?cite = 40.14.060

WebAIM. 2012. Screen Reader User Survey #4 Results. http://webaim.org/projects/screenreadersurvey4/#intro

WMATA. n.d. Washington Metropolitan Area Transit Authority Social Media Terms of Use. http://www.wmata.com/about_metro/metro_forward/pdf/SocialMedia_Policy.pdf

Zickuhr, K. and M. Madden. 2012. Older adults and internet use. Pew Internet & American Life Project. http://www.pewinternet.org/Reports/2012/Older-adults-and-internet-use.aspx

7

Tying It All Together

Kari Edison Watkins

There is little doubt that social media has revolutionized communications. Over the past decade, social media has served many transportation organizations by encouraging interaction with customers. It allowed the sharing of information, experiences, and opinions, as well as videos, photos, and even locations. Social media has allowed a substantial portion of the world's online population to interact, collaborate, and create connections.

Almost every state department of transportation has a social media presence, along with a multitude of transit agencies, more than 200 airports around the world, and other agencies at every level of government. Social media takes many forms and encompasses multiple platforms and channels:

- **Social and professional networking** sites are online platforms that support connections among members who may share personal or professional interests, activities, or backgrounds. Examples are Facebook, LinkedIn, GooglePlus, MySpace, GovLoop, and Ning.
- **Blogging** involves online journals that are updated on a regular basis. Individuals or organizations post news, commentaries, photographs, and media clips via platforms such as WordPress, Blogger, or Posterous.
- **Micro-blogging** is a form of blogging that allows users to post short updates and links to websites and online media such as Twitter and Tumblr.

281

- **Media- and document-sharing** sites are used by members to post and share video clips, photographs, documents, and presentations. The sites include YouTube, Flickr, Scribd, SlideShare, Vimeo, and GoogleDrive.
- **Social curation** sites encourage users to consolidate posts from multiple websites to illustrate or amplify a theme; examples are Pinterest and Storify.
- **Geolocation** applications enable users to share their locations with other members of their social network to provide information about places and services in the vicinity. Examples are Foursquare and location components of other sites.
- **Crowdsourcing** applications allow organizations to tap into the collective intelligence of the public or a defined group to solve a problem or prioritize issues, concerns, or recommendations.

This book has included examples from transportation agencies and advocacy groups at the forefront of social media, presenting their success stories and lessons learned.

USES OF SOCIAL MEDIA

Transportation organizations have incorporated social media into marketing, community activism, planning, real-time communications, and emergency management. Social media are powerful marketing tools, and agencies have used the 4 Es of social marketing to *entice* their customers to participate, *exchange* information to *engage* them in social media dialogues, and lead to a complete *experience* that leaves an overall impression. Successful marketing efforts such as contests engage customers in activities like naming a new farecard, promotions like Los Angeles Metro's tongue-in-cheek Miss Traffic campaign on YouTube, and Washington State DOT's Twitter-happy tunnel-boring machine.

In the realm of advocacy, social media provide easy-to-use and affordable tools that can help support grassroots campaigns and agency initiatives. Social media activities can reach existing supporters and hone the collective message. Advocates can use social outreach to organize citizens in order to pack hearing rooms, overwhelm the media with their presence, create noise about important issues, and ultimately achieve victories.

Social media provide extraordinary opportunities to share information with customers in real time, informing the public about transit service delays, traffic tie-ups, road closures, construction detours, and waiting times at agency offices (like the DMV). For all types of service disruptions and delays, social media can provide the critical pieces of information that travelers want: accurate predictions of delays, reasons for delays, potential alternative routes, and expected times of return to normal conditions. Many agencies have begun to incorporate social media tools into their emergency communication protocols as well, allowing those who communicate with the public to coordinate with first responders. This approach can apply to planned disruptions such as Los Angeles' Carmageddon project or natural disasters such as Hurricane Sandy.

BENEFITS OF SOCIAL MEDIA

Social media is all about connections. It allows agencies to connect directly with their consumers without relying on the press to get the facts straight. Through social media, agencies can tell their own stories with words and images using blogs, YouTube, and Flickr to share information. Social media helps agencies communicate at lightning speed, often responding to citizen requests or providing emergency information in a matter of minutes.

Social media can promote transparency for government agencies. Agencies can use social media to tell people what they do—how they spend public funds and make policy decisions—in order to help combat the distrust that comes from a lack of understanding. And they can use social media to reach out to citizens for feedback on projects; they can ask their customers for advice, and crowdsource solutions for local issues.

Social media can widen the reach of an agency. Agencies can use social tools to communicate with those who cannot attend public meetings and events by enabling them to obtain information and participate on their own schedules. In particular, social media brings younger residents who may not otherwise attend meetings into the fold. Although traditional outreach techniques such as public meetings, printed materials, e-mail, and project websites will remain critical components of public involvement for a long time to come, social media can complement these activities and allow agencies to connect with a larger audience. This is true whether a project is a subway expansion in Los Angeles or a long-range planning process in Austin.

ENTERING THE SOCIAL SPACE

Transportation organizations overwhelmingly emphasized the importance of having a presence in the social space. The emphasis of social media on engagement can be a scary concept for agencies that are used to one-way communications with their constituents. But successful social networkers advise these latecomers that not having an online presence can be worse. Others may jump in to fill the vacuum, possibly sending out the wrong message or incorrect information. Agencies should take the opportunities to break their own news, good or bad, and become the primary sources of information about their services and activities.

The public is already talking about the transportation services they use. An agency that is not engaged on social media loses the chance to be part of the conversation. Not engaging with followers can also have the negative result of giving the perception that an agency is not interested in its customers. By participating in social media, an agency has an opportunity to clarify inaccurate information, demonstrate its commitment to the community, and provide critical customer service.

The following sections synthesize best practices for transportation agency use of social media based on the experiences of transportation organizations in the social space.

Establish a Foundation

Before organizations jump into social media, they are well advised to lay the groundwork for a successful program by developing goals and objectives, defining staff roles, and making plans for evaluating progress. For many agencies, the first step is developing a social media policy. The policy does not have to be detailed or all-inclusive and can often be based on existing human resource and communications guidelines.

Some transportation organizations start by establishing protocols for their use of social media, including how they will use each platform and how they will manage content to balance agency needs with the spontaneous nature of social media (some require approval and others allow employee judgment). The second basic question is how to manage staff resources such as time commitment (average is one person week per month), frequency of monitoring (at least daily, often real time or hourly), and how to distribute relevant information within the agency to appropriate personnel. The third basic question is employee use of social media. The answer should define the numbers and types of employees who have permission to access social media sites and what sites they are authorized to visit. Finally, social media policies frequently address legal and institutional requirements such as archiving and record-keeping rules, protections for user privacy, and approaches for safeguarding the agency against cyber viruses and malware.

Find the Right Person

Organizations getting started must quickly decide who should post, tweet, and monitor their social media accounts. This is not as easy as it sounds. Being concise and quickly getting to the point in communicating with customers while still representing the goals of the agency are vital skills.

Most of the agencies interviewed for this book used a division of labor; one or two people approved the overall message and multiple contributors generated content for their social media channels. Pooling internal resources can create highly engaging content, but staff have to work together to ensure consistent messaging. An agency's posters should meet regularly to review successful and unsuccessful posts in order to improve technique and consistency.

When possible, a second party should review content before posting to identify materials that may be incorrect, inappropriate, or just misguided. As one interviewee recommended, "Think twice before you tweet things. Count to ten." Once an organization finds the right person, however, that individual should be given some leeway to post without a lengthy review process. Otherwise, the content will not take advantage of the critical real-time nature of social media.

Find the Right Voice

Each agency will have its own voice but the language of social media is almost always informal. Everyone posting on behalf of an agency should make sure that content reflects the branding of the organization and builds upon it. Above all, social media interactions should be personal rather than automated. Automated messages tend to be dry and fail to engage their audiences. If the agency's posts show personality and include some fun, people will relate to the person behind the posts and the audience will build.

Use the Right Platform

Each social media platform has differing strengths and reaches different audiences. Agencies should match their message to the right platform. Some agencies may use Twitter for real-time alerts but find that Facebook serves them better for customer engagement. Others prefer to share organizational information through blogs or post instructional videos on YouTube. Regardless of the platform, the messages should be consistent and integrated with the agency's overall communications plan.

Develop the Right Content

Finding good content is a strategic endeavor. To be successful, social media specialists pay attention to pop culture, know their audiences, and use a conversational—even humorous—tone. Good content cannot always be planned; sometimes the best posts come from an immediate reaction to an unplanned event, whether a hurricane, accident, or lost-and-found teddy bear.

Many agencies mix news releases with real-time information and special-interest stories; they often share appropriate links to other organizations' social media. Live-tweeting meetings, presentations, and hearings can open such events to a wider audience. When an agency has advance knowledge of the content (for example, for a planned presentation), the posts can be set up in advance so that the tweeter can frame the story ahead of time and capture the meeting's energy in real time. Between events, the content creator can solicit questions from the community for managers of the organization to answer.

The key is to provide correct, concise, and clear information that citizens want to know, while also providing them with information the agency wants them to receive. Overall, agencies must be transparent, accountable, and available. They must be there to listen, let people know that they are listening, and hear what their customers say. Transparency also means that agencies must be willing to look honestly at their own programs, criticize their own actions when warranted, and apologize for their mistakes when necessary.

It goes without saying that accuracy is important; the viral nature of social media means that inaccuracies can be rebroadcast with little or no control by an agency. The information must be concise because social media constitutes a very crowded platform and it is getting harder for any one agency to break through with its message. Lengthening or overposting a message does not mean it will reach more people; it means people will tire and turn to other channels. Some agencies believe that having more than one account can splinter their audiences and make it difficult to create good content on all sites; others develop specialized accounts for different services.

Maintain the Audience

Building an audience and maintaining it require time and consistency. Often emergency situations will create surges in followers; the challenge for an agency is to maintain that audience after the emergency has been resolved. Once an agency starts posting, people expect the information flow to continue. Agencies can keep their audiences by engaging them often.

   Commitment is one of the keys to a successful social media program. Social media tools may be free and easy to use, but agencies often underestimate the time required to find and post content, monitor varying platforms, and measure the impact of a program. Monitoring and responding to social media posts and comments can be a 24-hour job; peak activity often occurs in the evening hours when people are home from work. To communicate real-time situations like roadway closures, agencies often send updates multiple times per day, seven days per week. Emergencies may require communications around the clock. It may not be possible for public agencies to routinely staff their social media accounts during non-business hours; therefore they should let their audiences know when their accounts are monitored.

Address Criticism

Truly engaging an audience requires two-way communication between an agency and its followers. Organizations interviewed for this book overwhelmingly found that they received more positive than negative comments, but inevitably something will go wrong. People may post inaccurate information about an agency or its services or criticize its policies. Savvy agencies quickly correct misinformation and many attempt to reach out to disgruntled citizens through one-on-one conversations. Some naysayers will never be convinced, but this kind of personal connection has helped some agencies turn vocal critics into staunch advocates.

Ensure Information Equity

Social media is delivering new audiences to many public agencies. To ensure that everyone can access their online information, agencies must use appropriate practices and technologies to reach all audiences. Agencies should make every effort to ensure that citizens with sensory or cognitive impairments can access their social media content. Ensuring that images and videos are captioned and structuring posts to allow use by screen readers can help make posts, tweets, and shared media accessible to all citizens.

When appropriate, agencies should also account for people who do not speak English as a first language and provide ways to allow their blogs, tweets, and posts to be read in other languages. Such efforts help maintain the open and accessible nature of social media.

Measure the Results

Government agencies will want to measure the impacts of their social media programs. Measuring and sharing successes internally can build support within an agency and ultimately justify current and future program investment. Without measuring the impact of social media, an agency will not be able to answer questions about the return on investment for its social activities. Fortunately, checking the pulse of social media is easy. Many baseline metrics and analytics are readily available, simple to use, and often free.

Ideally, an agency should establish a measurement program from the beginning that follows the overall objectives of the program and evaluates progress toward reaching the objectives. A good measurement program will be both quantitative and qualitative in nature. Quantitative measures should reveal the numbers of followers and friends and also the numbers of comments, retweets, and shares. Qualitative measures cover engagement and influence by rating the tone (positive, negative, neutral) and reach (broad, moderate, narrow) of responses.

An agency should create a consistent and ongoing measurement program. It should define a particular time each week for measurement activity via automatic web reports (GoogleAnalytics, FacebookInsights, HootSuite), consistent scoring, and generation of easy-to-use standardized reports that can track activities over time.

Keep Social Media in Context

Agencies must remember that social media will never reach everyone. Social media is a powerful tool, but only one type among many available to transportation organizations. Emergency communications require having a plan and an internal communication network in place for rapidly evolving situations. Similarly, a system should be in place to convey official service requests to the appropriate department for response and tracking. Social media should augment communications but not replace traditional outreach methods.

THE FUTURE

Social media will continue to be used to increase public engagement and disseminate agency messages and critical information in real time. In addition, social media analysis tools are pushing transportation agencies into a new age that allows frequent and occasional customers a greater say and more positive impact on their systems. Social computing or crowdsourcing, social media-based surveying, and data mining and activity pattern recognition are evolving as means to reach respondents, understand their social networks and activities, and analyze their opinions to improve transportation.

Social computing, in which crowds of people collaborate with computer systems to perform tasks neither could accomplish alone, is improving the efficiency and effectiveness of citizen participation. Social media allow agencies to tap into large numbers of participants to engage in activities such as citizen science (people working as sensors to relay information), crowdsourcing (people working together to identify a system's assets and issues), human computation (people as informational resources for computational systems), and participatory sensing (mobile phones as instruments and sensors).

Applications such as Waze, Tiramisu, and Roadify use the knowledge of the people to identify transportation tie-ups or delays. ShareAbouts and Urban Mediator allow ranking and discussion to locate amenities. SeeClickFix, FixMyStreet, and ParkScan encourage the identification of infrastructure issues. Twitter is used as a source for gathering real-time data from populations during evacuations and public planning. These tools even go a step further to engage end users to design services in conjunction with providers. This co-design process is intended to create services that are more responsive to the changing needs of citizens, foster trust in government, and increase social capital and sense of community.

The survey research industry is also working to incorporate the benefits of mobile technology and social media into survey techniques. It is currently impossible to conduct a randomly sampled quantitative survey via social media due to biases in social media use. However, researchers see great opportunities to leverage the power of social media to promote survey opportunities, test survey questions, and supplement traditional surveys. Agencies can use social media to promote online surveys directly or recruit participants for panels of interested customers who can be polled on a variety of topics over a set period. The recruitment message can tell users which routes or corridors will be surveyed, summarize the purpose and value of a survey, and encourage users to participate. To test survey questions, organizations can share the wording of specific questions or run an online survey to ensure that participants can understand the survey language and instructions. In addition, marketing messages and other aspects of survey design can be tested. Supplemental measures such as non-random qualitative surveys can take advantage of the viral nature of social media by encouraging customers and partners to post or retweet links to general public surveys. Finally, researchers are moving toward capturing online social network characteristics to supplement traditional survey-based transportation studies.

The wealth of information social media provides can be used in observing patterns to understand people's behaviors, choices, and opinions. Status updates, media sharing, and geo-located check-ins can tell researchers what is important, how people feel, what they are doing, and where they travel. By using social media, analysts can visualize urban dynamics, understand the underlying activity participation, location choice, and sentiment, and even influence the patterns.

Examples of issues to be resolved within this data source include concerns about privacy, self-selection of social media users, and a lack of detailed or complete information. Although concepts of privacy are changing over time, the trust that users provide will remain only if they are not left vulnerable to malicious activities when they provide information about their movements and activities.

Social media usage is spreading, but care must still be taken to ensure that the inherent self-selection does not result in poor analysis due to bias. Finally, social media data is becoming richer and more complete, but further research is needed to obtain key information about socio-economic characteristics, activity duration, and complete daily patterns before these media can become true sources of transportation information.

At the simplest level, agencies can use data mining of social media to understand public perceptions and satisfaction based on polarity and classification of posts. Looking toward the future, real-time geo-located social media posts can be used as sensors to observe and understand the behaviors of millions of people to model and influence transportation system performance.

CONCLUSIONS

Although social media will continue to evolve, its presence has already been felt worldwide. Social networking, blogging and micro-blogging, online media sharing, social curation, and even geolocation and crowd-sourcing have exerted profound impacts on the transportation industry.

Agencies that entered the social media realm have used it to increase their reach, engage their customers, and provide accurate real-time information. Although an agency should monitor policies, content, and return on investment carefully, with proper guidance and attention to lessons learned, the benefits of social media usage can far outweigh the costs.

INDEX